Nanostructured Materials for Energy Storage and Conversion

Nanostructured Materials for Energy Storage and Conversion

Editor

Luca Pasquini

MDPI • Basel • Beijing • Wuhan • Barcelona • Belgrade • Manchester • Tokyo • Cluj • Tianjin

Editor
Luca Pasquini
Department of Physics and
Astronomy
University of Bologna
Bologna
Italy

Editorial Office
MDPI
St. Alban-Anlage 66
4052 Basel, Switzerland

This is a reprint of articles from the Special Issue published online in the open access journal *Nanomaterials* (ISSN 2079-4991) (available at: www.mdpi.com/journal/nanomaterials/special_issues/nanostructured_energy_conversion).

For citation purposes, cite each article independently as indicated on the article page online and as indicated below:

LastName, A.A.; LastName, B.B.; LastName, C.C. Article Title. *Journal Name* **Year**, *Volume Number*, Page Range.

ISBN 978-3-0365-4184-6 (Hbk)
ISBN 978-3-0365-4183-9 (PDF)

© 2022 by the authors. Articles in this book are Open Access and distributed under the Creative Commons Attribution (CC BY) license, which allows users to download, copy and build upon published articles, as long as the author and publisher are properly credited, which ensures maximum dissemination and a wider impact of our publications.

The book as a whole is distributed by MDPI under the terms and conditions of the Creative Commons license CC BY-NC-ND.

Contents

Luca Pasquini
Nanostructured Materials for Energy Storage and Conversion
Reprinted from: *Nanomaterials* **2022**, *12*, 1583, doi:10.3390/nano12091583 1

Liyufen Dai, Xiangli Zhong, Juan Zou, Bi Fu, Yong Su and Chuanlai Ren et al.
Highly Ordered SnO_2 Nanopillar Array as Binder-Free Anodes for Long-Life and High-Rate Li-Ion Batteries
Reprinted from: *Nanomaterials* **2021**, *11*, 1307, doi:10.3390/nano11051307 5

Tahar Azib, Claire Thaury, Fermin Cuevas, Eric Leroy, Christian Jordy and Nicolas Marx et al.
Impact of Surface Chemistry of Silicon Nanoparticles on the Structural and Electrochemical Properties of $Si/Ni_{3.4}Sn_4$ Composite Anode for Li-Ion Batteries
Reprinted from: *Nanomaterials* **2020**, *11*, 18, doi:10.3390/nano11010018 15

Timotheus Jahnke, Leila Raafat, Daniel Hotz, Andrea Knöller, Achim Max Diem and Joachim Bill et al.
Highly Porous Free-Standing rGO/SnO_2 Pseudocapacitive Cathodes for High-Rate and Long-Cycling Al-Ion Batteries
Reprinted from: *Nanomaterials* **2020**, *10*, 2024, doi:10.3390/nano10102024 29

Bishal Kafle, Ahmed Ismail Ridoy, Eleni Miethig, Laurent Clochard, Edward Duffy and Marc Hofmann et al.
On the Formation of Black Silicon Features by Plasma-Less Etching of Silicon in Molecular Fluorine Gas
Reprinted from: *Nanomaterials* **2020**, *10*, 2214, doi:10.3390/nano10112214 47

Suzan Saber, Bernabé Marí, Andreu Andrio, Jorge Escorihuela, Nagwa Khattab and Ali Eid et al.
Structural and Electrochemical Analysis of CIGS: Cr Crystalline Nanopowders and Thin Films Deposited onto ITO Substrates
Reprinted from: *Nanomaterials* **2021**, *11*, 1093, doi:10.3390/nano11051093 63

Bumjin Gil, Jinhyun Kim, Alan Jiwan Yun, Kimin Park, Jaemin Cho and Minjun Park et al.
$CuCrO_2$ Nanoparticles Incorporated into PTAA as a Hole Transport Layer for 85 °C and Light Stabilities in Perovskite Solar Cells
Reprinted from: *Nanomaterials* **2020**, *10*, 1669, doi:10.3390/nano10091669 81

Radenka Krsmanović Whiffen, Amelia Montone, Loris Pietrelli and Luciano Pilloni
On Tailoring Co-Precipitation Synthesis to Maximize Production Yield of Nanocrystalline Wurtzite ZnS
Reprinted from: *Nanomaterials* **2021**, *11*, 715, doi:10.3390/nano11030715 95

Marco Calizzi, Robin Mutschler, Nicola Patelli, Andrea Migliori, Kun Zhao and Luca Pasquini et al.
CO_2 Hydrogenation over Unsupported Fe-Co Nanoalloy Catalysts
Reprinted from: *Nanomaterials* **2020**, *10*, 1360, doi:10.3390/nano10071360 107

Michele Mazzanti, Stefano Caramori, Marco Fogagnolo, Vito Cristino and Alessandra Molinari
Turning Waste into Useful Products by Photocatalysis with Nanocrystalline TiO_2 Thin Films: Reductive Cleavage of Azo Bond in the Presence of Aqueous Formate
Reprinted from: *Nanomaterials* **2020**, *10*, 2147, doi:10.3390/nano10112147 119

Ruben F. Hamans, Rifat Kamarudheen and Andrea Baldi
Single particle approaches to energy conversion in plasmonic catalysts
Reprinted from: *Nanomaterials* **2020**, *10*, 2377, doi:10.3390/nano10122377 . . . . **137**

Editorial

Nanostructured Materials for Energy Storage and Conversion

Luca Pasquini

Department of Physics and Astronomy, University of Bologna, Viale C. Berti-Pichat 6/2, 40127 Bologna, BO, Italy; luca.pasquini@unibo.it

The conversion and storage of renewable energy sources is an urgent challenge that we need to tackle to transition from a fossil fuel-based economy to a low-carbon society. I can hardly imagine that this revolution could take place without further breakthroughs in materials science and technology. Indeed, contemporary materials history highlights many game-changing materials that have deeply impacted our lives and contributed to a reduction in carbon dioxide emissions. High-efficiency photovoltaic cells, blue light-emitting diodes, and cathodes for Li-ion batteries are among the most illuminating examples of knowledge-based materials' development, which have experienced an exponential market permeation and received the highest scientific awards.

These success stories, like many others in materials science, were built upon a tailored control of the interconnected processes that take place at the nanoscale, such as charge excitation, charge transport and recombination, ionic diffusion, intercalation, and the interfacial transfer of matter and charge. Nanostructured materials, thanks to their ultra-small building blocks and the high interface-to volume-ratio, offer a rich toolbox to the scientist that aspires to boost the energy conversion efficiency or the power and energy density of a material. Examples of the materials' tailoring tools enabled by nanoscience include: (i) the quick separation and collection of photoexcited charges, avoiding recombination issues; (ii) high catalytic activity thanks to the extreme surface area; (iii) accelerated diffusion of ions and atoms along the nanocrystallite interfaces, and (iv) enhanced light harvesting due to the low reflectivity of nanostructured surfaces. Furthermore, new phenomena will arise in nanoparticles (NPs), such as the surface plasmon resonance, which dramatically alters the interaction between metals and the electromagnetic field, superparamagnetism, which turns a ferromagnetic particle into a collective paramagnet, and exciton confinement, which causes the size-dependent colour of semiconductor quantum dots.

The ten articles published in this Special Issue showcase the different applications of nanomaterials in the field of energy storage and conversion, including electrodes for Li-ion batteries (LIBs) and beyond [1–3], photovoltaic materials [4–6], pyroelectric energy harvesting [7], and (photo)catalytic processes [8–10]. The scientific contributions are briefly summarized in the following.

Three main types of anode materials are currently being investigated for the replacement of graphite in LIBs: (i) novel carbonaceous materials, (ii) conversion-type transition metal compounds, and (iii) Si- and Sn-based anodes. Dai et al. report on the electrochemical properties of an ordered array of SnO_2 nanopillars prepared by pulsed laser deposition on a nanoporous alumina template and employed as conversion-type anode for LIBs [1]. The ordered nanopillar architecture provides adequate room for volumetric expansion during lithiation/delithiation, offering a strategy to mitigate the performance degradation that affects conversion-type anodes. The improved structural integrity and stability allows for a high specific capacity of 524/313 mAh/g to be maintained after 1100/6500 cycles. In the work of Azib et al., the surface chemistry of Si nanoparticles in a $Si/Ni_{3.4}Sn_4$ composite anode is modified by a coating of either carbon or oxide [2]. The coating strongly reduces the reaction between Si and $Ni_{3.4}Sn_4$ during the composite's preparation by ball milling. Better lithiation properties are obtained for carbon-coated Si particles that can deliver over

Citation: Pasquini, L. Nanostructured Materials for Energy Storage and Conversion. *Nanomaterials* **2022**, *12*, 1583. https://doi.org/10.3390/nano12091583

Received: 1 May 2022
Accepted: 5 May 2022
Published: 7 May 2022

Publisher's Note: MDPI stays neutral with regard to jurisdictional claims in published maps and institutional affiliations.

Copyright: © 2022 by the author. Licensee MDPI, Basel, Switzerland. This article is an open access article distributed under the terms and conditions of the Creative Commons Attribution (CC BY) license (https://creativecommons.org/licenses/by/4.0/).

500 mAh/g for at least 400 cycles. Jahnke at al. studied a highly porous aerogel cathode composed of reduced graphene oxide (rGO), which is loaded with nanostructured SnO_2 that serves as a cathode material for high-rate Al-ion batteries [3]. This binder-free hybrid has excellent mechanical properties and combines the pseudocapacity of rGO with the electrochemical capacity of SnO_2 nanoplatelets. The proposed design is appealing for future energy storage devices that can accommodate ionic species other than Li^+.

The formation of nanoscale features with a high aspect ratio and large surface area on Si is of great interest not only for Si-based battery anodes but also for photovoltaic applications, where the ability to reduce surface reflection leads to enhanced solar harvesting. Kafle et al. report on the formation of such anti-reflective nanostructures, known as "black silicon" features, by the plasma-less etching of crystalline Si using F_2 gas [4]. This approach may provide an industrial etching tool for products that require a high-volume manufacturing platform. In the field of photovoltaics, Saber et al. present a new approach to the synthesis of nanopowders and thin films of a $CuInGaSe_2$ (CIGS) chalcopyrite material doped with different amounts of Cr [5]. From electrochemical impedance spectroscopy measurements, they conclude that thin films with a $CuIn_{0.4}Cr_{0.2}Ga_{0.4}Se_2$ composition are promising materials for solar cell applications. Gil et al. report on the synthesis and properties of a hybrid hole transport layer for perovskite photovoltaics, in which high-mobility $CuCrO_2$ nanoparticles are incorporated into conventional PTAA. This approach can boost hole extraction while passivating deep-level traps. Moreover, a stability of about 900 h is achieved under 85 °C, 85% relative humidity and continuous 1 sun illumination, suggesting an effective strategy to improve the durability of perovskite solar cells. Pyroelectric energy harvesting, which exploits the temperature changes induced by industrial activity or occurring naturally, can also contribute to the clean energy transition. The work by Krsmanović Whiffen et al. describes a co-precipitation method for the synthesis of nanocrystalline wurtzite ZnS, which is an interesting material for pyroelectric applications [7].

Three papers in this Special Issue focus on thermocatalysis, photocatalysis, and plasmon-driven catalysis. Calizzi et al. studied the thermocatalytic hydrogenation of CO_2 by Fe-Co alloy nanoparticles with different compositions, as prepared by inert gas condensation [8]. The Fe-Co nanoalloys can catalyse the formation of C_2-C_5 hydrocarbons, which are not detected using elemental Co and Fe nanoparticles. This effect is attributed to the simultaneous variations in CO_2 binding energy and decomposition barrier as a function of the Fe/Co ratio in the nanoalloy. Mazzanti et al. report on the reductive cleavage of azo bonds by the UV photoexcitation of nanostructured TiO_2 films in contact with an aqueous solution of azo dyes [9]. Charge separation is extremely long-lived in nanostructured TiO_2 thin films, making them suitable for driving both oxidation and reduction reactions. This approach provides an effective solution for the simultaneous implementation of wastewater purification and photocatalytic conversion of waste into useful products. Finally, plasmonic nanoparticles have recently attracted interest in the field of photocatalysis thanks to their ability to harvest and convert light into highly energetic charge carriers and heat. The perspective article written by Hamans et al. highlights two techniques for single-particle studies of structure–function relations between plasmonic nanocatalysts: surface-enhanced Raman spectroscopy and super-resolution fluorescence microscopy [10]. These two far-field optical techniques make it possible to take a closer look at fundamental nanoscale processes, such as photoactivation mechanism, molecular intermediates, and reaction pathways.

I wish to express my deepest gratitude to all the authors who contributed to this Special Issue and to the reviewers who generously dedicated their time to improving the quality of the submitted manuscripts. Special thanks go to all the editorial staff of *Nanomaterials* for their enduring support during the preparation and publication of this Special Issue.

Funding: This research received no external funding.

Conflicts of Interest: The author declares no conflict of interest.

References

1. Dai, L.; Zhong, X.; Zou, J.; Fu, B.; Su, Y.; Ren, C.; Wang, J.; Zhong, G. Highly Ordered SnO_2 Nanopillar Array as Binder-Free Anodes for Long-Life and High-Rate Li-Ion Batteries. *Nanomaterials* **2021**, *11*, 1307. [CrossRef] [PubMed]
2. Azib, T.; Thaury, C.; Cuevas, F.; Leroy, E.; Jordy, C.; Marx, N.; Latroche, M. Impact of Surface Chemistry of Silicon Nanoparticles on the Structural and Electrochemical Properties of Si/Ni3.4Sn4 Composite Anode for Li-Ion Batteries. *Nanomaterials* **2020**, *11*, 18. [CrossRef] [PubMed]
3. Jahnke, T.; Raafat, L.; Hotz, D.; Knöller, A.; Diem, A.M.; Bill, J.; Burghard, Z. Highly Porous Free-Standing rGO/SnO2 Pseudocapacitive Cathodes for High-Rate and Long-Cycling Al-Ion Batteries. *Nanomaterials* **2020**, *10*, 2024. [CrossRef] [PubMed]
4. Kafle, B.; Ridoy, A.I.; Miethig, E.; Clochard, L.; Duffy, E.; Hofmann, M.; Rentsch, J. On the Formation of Black Silicon Features by Plasma-Less Etching of Silicon in Molecular Fluorine Gas. *Nanomaterials* **2020**, *10*, 2214. [CrossRef] [PubMed]
5. Saber, S.; Marí, B.; Andrio, A.; Escorihuela, J.; Khattab, N.; Eid, A.; El Nahrawy, A.; Aly, M.A.; Compañ, V. Structural and Electrochemical Analysis of CIGS: Cr Crystalline Nanopowders and Thin Films Deposited onto ITO Substrates. *Nanomaterials* **2021**, *11*, 1093. [CrossRef] [PubMed]
6. Gil, B.; Kim, J.; Yun, A.J.; Park, K.; Cho, J.; Park, M.; Park, B. $CuCrO_2$ Nanoparticles Incorporated into PTAA as a Hole Transport Layer for 85 °C and Light Stabilities in Perovskite Solar Cells. *Nanomaterials* **2020**, *10*, 1669. [CrossRef] [PubMed]
7. Krsmanović Whiffen, R.; Montone, A.; Pietrelli, L.; Pilloni, L. On Tailoring Co-Precipitation Synthesis to Maximize Production Yield of Nanocrystalline Wurtzite ZnS. *Nanomaterials* **2021**, *11*, 715. [CrossRef] [PubMed]
8. Calizzi, M.; Mutschler, R.; Patelli, N.; Migliori, A.; Zhao, K.; Pasquini, L.; Züttel, A. CO_2 Hydrogenation over Unsupported Fe-Co Nanoalloy Catalysts. *Nanomaterials* **2020**, *10*, 1360. [CrossRef] [PubMed]
9. Mazzanti, M.; Caramori, S.; Fogagnolo, M.; Cristino, V.; Molinari, A. Turning Waste into Useful Products by Photocatalysis with Nanocrystalline TiO_2 Thin Films: Reductive Cleavage of Azo Bond in the Presence of Aqueous Formate. *Nanomaterials* **2020**, *10*, 2147. [CrossRef] [PubMed]
10. Hamans, R.F.; Kamarudheen, R.; Baldi, A. Single particle approaches to plasmon-driven catalysis. *Nanomaterials* **2020**, *10*, 2377. [CrossRef] [PubMed]

Article

Highly Ordered SnO₂ Nanopillar Array as Binder-Free Anodes for Long-Life and High-Rate Li-Ion Batteries

Liyufen Dai [1,2], Xiangli Zhong [2], Juan Zou [1,2], Bi Fu [1], Yong Su [2], Chuanlai Ren [1], Jinbin Wang [2] and Gaokuo Zhong [1,*]

[1] Shenzhen Key Laboratory of Nanobiomechanics, Shenzhen Institutes of Advanced Technology, Chinese Academy of Sciences, Shenzhen 518055, China; lyf.dai@siat.ac.cn (L.D.); zjuan@xtu.edu.cn (J.Z.); fub@sustech.edu.cn (B.F.); cl.ren@siat.ac.cn (C.R.)

[2] School of Materials Science and Engineering, Xiangtan University, Xiangtan 411105, China; xlzhong@xtu.edu.cn (X.Z.); susu0324@163.com (Y.S.); jbwang@xtu.edu.cn (J.W.)

* Correspondence: gk.zhong@siat.ac.cn

Citation: Dai, L.; Zhong, X.; Zou, J.; Fu, B.; Su, Y.; Ren, C.; Wang, J.; Zhong, G. Highly Ordered SnO₂ Nanopillar Array as Binder-Free Anodes for Long-Life and High-Rate Li-Ion Batteries. *Nanomaterials* **2021**, *11*, 1307. https://doi.org/10.3390/nano11051307

Academic Editors: Luca Pasquini and Sergio Brutti

Received: 17 March 2021
Accepted: 7 May 2021
Published: 15 May 2021

Publisher's Note: MDPI stays neutral with regard to jurisdictional claims in published maps and institutional affiliations.

Copyright: © 2021 by the authors. Licensee MDPI, Basel, Switzerland. This article is an open access article distributed under the terms and conditions of the Creative Commons Attribution (CC BY) license (https://creativecommons.org/licenses/by/4.0/).

Abstract: SnO₂, a typical transition metal oxide, is a promising conversion-type electrode material with an ultrahigh theoretical specific capacity of 1494 mAh g^{-1}. Nevertheless, the electrochemical performance of SnO₂ electrode is limited by large volumetric changes (~300%) during the charge/discharge process, leading to rapid capacity decay, poor cyclic performance, and inferior rate capability. In order to overcome these bottlenecks, we develop highly ordered SnO₂ nanopillar array as binder-free anodes for LIBs, which are realized by anodic aluminum oxide-assisted pulsed laser deposition. The as-synthesized SnO₂ nanopillar exhibit an ultrahigh initial specific capacity of 1082 mAh g^{-1} and maintain a high specific capacity of 524/313 mAh g^{-1} after 1100/6500 cycles, outperforming SnO₂ thin film-based anodes and other reported binder-free SnO₂ anodes. Moreover, SnO₂ nanopillar demonstrate excellent rate performance under high current density of 64 C (1 C = 782 mA g^{-1}), delivering a specific capacity of 278 mAh g^{-1}, which can be restored to 670 mAh g^{-1} after high-rate cycling. The superior electrochemical performance of SnO₂ nanoarray can be attributed to the unique architecture of SnO₂, where highly ordered SnO₂ nanopillar array provided adequate room for volumetric expansion and ensured structural integrity during the lithiation/delithiation process. The current study presents an effective approach to mitigate the inferior cyclic performance of SnO₂-based electrodes, offering a realistic prospect for its applications as next-generation energy storage devices.

Keywords: lithium-ion batteries; SnO₂; nanoarray; anode; high-rate

1. Introduction

Rechargeable lithium-ion batteries (LIBs) enjoy superior energy density and high portability, realizing their widespread utilization in our commonly used electronic devices, such as mobile phones, cameras, and laptops [1]. In the last decade, the utilization of rechargeable LIBs has rapidly expanded into the field of electric vehicles (EVs) [2]. The development of next-generation EVs requires LIBs with excellent energy density and rapid charge/discharge capability under high current densities. Currently, the commercialization of LIBs is mainly based on the carbon anodes with graphitic layered structure [3], however, the limited theoretical capacity (372 mAh g^{-1}) of graphite cannot meet the requirements of next-generation LIBs, driving the exploration of alternative anode materials with high Li storage capacity. Among a wide array of anode materials, transition metal oxides (TMOs) are considered promising candidates against commercial graphite due to its rich in natural resources and outstanding Li storage capacity [4,5]. In general, the Li-storage mechanism in TMOs is either intercalation/deintercalation ($M_xO_y + nLi^+ + ne^- \leftrightarrow Li_nM_xO_y$) or conversion reaction ($M_xO_y + 2yLi^+ + 2ye^- \leftrightarrow yLi_2O + xM$) [6]. One should note that the conversion-type TMOs render high theoretical Li storage capacity, such as Fe₃O₄

(926 mAh g^{-1}) [7], Co_3O_4 (890 mAh g^{-1}) [8], and SnO_2 (1494 mAh g^{-1}) [9]. However, the conversion-type TMOs experience large volumetric change during the charge/discharge process, e.g., the volume change of SnO_2 by ~300% [10,11], resulting in rapid capacity decay, inferior cyclic performance, and poor rate capability.

After the first report on the utilization of SnO_2 as an anode in LIBs by Idota et al. in 1997 [12], extensive research has been carried out to solve the problem of volumetric expansion [13–15]. Benefiting from the development of nanotechnology and nanoscience [16–18], a wide variety of SnO_2 nanostructures, such as nanorods [19], nanowires [20,21], nanotubes [22], and nanofibers [23], have been employed to improve the electrochemical performance and cyclic stability of SnO_2-based anodes. Nevertheless, these nanostructures of SnO_2 suffer from the problems of agglomeration, redundant interfaces, and limited electron transfer [24]. Hence, a highly ordered and stable SnO_2 nanostructure is desired to overcome these problems. Herein, we have fabricated highly arrayed SnO_2 nanopillar by anodic aluminum oxide (AAO)-assisted pulsed laser deposition (PLD) and employed as an anode electrode in LIBs. Benefiting from the unique nanoarray structure, the SnO_2 anode rendered a reversible specific capacity of about 830 mAh g^{-1} after 300 charge/discharge cycles and maintained a specific capacity of 313 mAh g^{-1} after 6500 charge/discharge cycles at the current density of 2 C, exhibiting superior Li storage capacity and excellent cyclic life. Note that 1 C indicates the current strength of the battery when it is fully discharged in one hour, and 1 C of SnO_2 is 782 mA g^{-1}. Furthermore, it shows the discharge capacity of 433, 414, 354, and 278 mAh g^{-1} at the current density of 8, 16, 32, and 64 C, respectively, indicates the ultrahigh rate capability of as-synthesized SnO_2 nanopillar. These results reveal the superior electrochemical performance of highly arrayed SnO_2 nanopillar, which can be employed as binder-free electrodes in next-generation energy storage devices.

2. Materials and Methods

2.1. SnO_2 Nanopillar Array Deposition

In order to improve the hydrophilicity of Cu foil surface and the adhesion between the AAO template and Cu foil, a plasma cleaning system was used to clean the surface of Cu foil before the transfer of AAO template. The employed AAO templates were purchased from Topmembranes Technology Co., LTD, and their thickness and aperture are around 650 nm and 310 nm, respectively. The fabrication of SnO_2 nanopillar array included three steps: (1) transfer of AAO template, (2) SnO_2 deposition, and (3) removal of AAO template. AAO-covered Cu foil and bare Cu foil were placed in a PLD system (Pascal Mobile Combi-Laser MBE) to deposit SnO_2. The optimal deposition conditions were achieved by setting the laser energy at 350 mJ, laser pulse frequency at 10 Hz, substrate temperature at 500 °C and oxygen pressure at 2.0×10^{-6} Torr. The removal of the AAO template was accomplished by using polyimide high-temperature adhesive tape.

2.2. Electrochemical Characterization

The SnO_2 nanoarray and thin-film electrodes (0.5 cm × 0.5 cm) with a mass loading of 0.12 mg/cm^2 and 0.2 mg/cm^2, respectively, were used for electrochemical characterization. The CR2025-type coin-cells were assembled in an Ar-filled gloves box with water and oxygen content of < 0.2 ppm. The 1 M electrolyte was prepared by adding an appropriate amount of $LiPF_6$ to the mixed solution of ethylene carbonate (EC) and dimethyl carbonate (DMC) (1:1 v/v). The involved separator material is PP (polypropylene) /PE (polyethylene) /PP from Guang-dong Canrd New Energy Technology Co., Ltd (Dongguan, China). The galvanostatic charge/discharge cycling and rate capability were carried out by using a battery testing system (2001A, LAND, Wuhan, China) in the voltage range of 0.01 to 3.0 V (vs. Li/Li$^+$). Cyclic voltammetry (CV) test was carried out by using an electrochemical workstation (CS2350, Wuhan, China).

3. Results

A highly arrayed SnO$_2$ nanopillar architecture has been designed to overcome large volumetric change of SnO$_2$ (~300%) during the charge/discharge process. As schematically illustrated in Figure 1a, the fabrication process of SnO$_2$ array includes three steps: (1) AAO template transfer, (2) SnO$_2$ deposition, and (3) removal of AAO template. First, the AAO template is carefully transferred on Cu foil before SnO$_2$ deposition and then, the SnO$_2$ array is deposited on Cu foil by PLD using the AAO template. Finally, an array of highly ordered SnO$_2$ nanopillar is left on the Cu substrate after the removal of AAO template. The utilization of SnO$_2$ nanoarray as an anode material in LIBs is schematically presented in Figure 1b, where the Li-ion half-cell consists of lithium metal, the separator, and SnO$_2$ nanoarray. One should note that the highly arrayed SnO$_2$ nanopillar can provide enough room for volumetric change, abundant active sites for Li storage, and ordered electron transmission channels. Furthermore, the architecture of SnO$_2$ array is confirmed by scanning electronic microscopy (SEM), as shown in Figure 1c,d. The top-view SEM image in Figure 1c clearly demonstrates the morphology of AAO template before and after SnO$_2$ deposition, revealing the well-maintained AAO template structure during PLD. Moreover, the SEM image in Figure 1d confirms the array architecture of the SnO$_2$ nanopillar, wherein the diameter of SnO$_2$ nanopillar and the gap between neighboring SnO$_2$ nanopillar can be estimated as ~300 nm and ~130 nm, respectively, which are consistent with the structure of AAO template. Note that SnO$_2$ nanopillar with different diameter and gap were employed before we use the optimized diameter and gap as ~300 nm and ~130 nm, because we found that too small an array gap and the large nanopillar were unable to achieve long term charge/discharge cycling, which may cause by they unable to provide enough space to overcome the large volumetric change during the charge/discharge process.

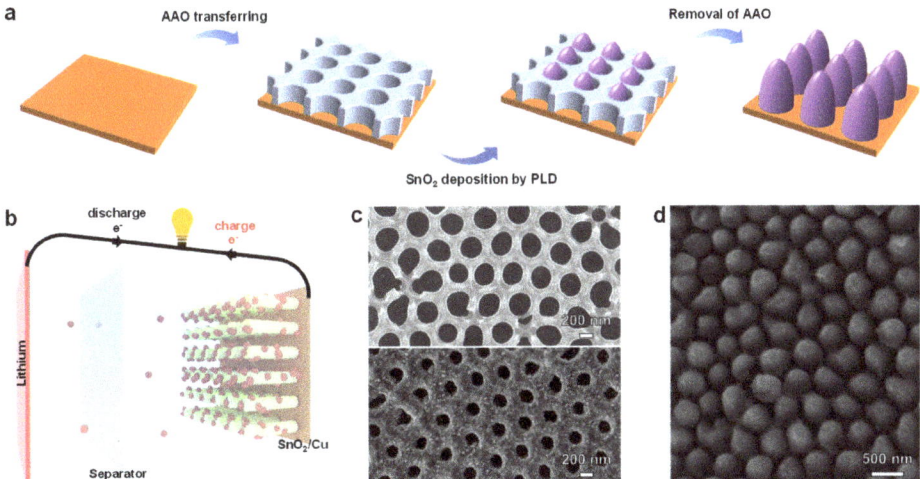

Figure 1. SnO$_2$ array deposition. (**a**) Fabrication of SnO$_2$ nanopillar array by PLD using AAO template on a Cu foil substrate; (**b**) Schematic diagram of a half-cell Li-ion battery, where SnO$_2$ nanoarray is used as an anode and Li-foil is used as a cathode and counter electrode; (**c**) SEM images of bare AAO template (upper panel) and AAO template after SnO$_2$ deposition (lower panel); and (**d**) a top-view SEM image of the SnO$_2$ nanopillar array after the removal of AAO template.

The X-ray diffraction (XRD) patterns of bare Cu foil and SnO$_2$-coated Cu foil are presented in Figure 2a. The diffraction peaks can be well-indexed to tetragonal rutile-like SnO$_2$ structure without any impurities [25]. The chemical composition of SnO$_2$ nanoarray is further studied by X-ray photoelectron spectroscopy (XPS). The wide-range XPS spectrum is presented in Figure 2b, showing the presence of Sn, O and Cu elements. Note that the silver (Ag) peak appeared due to the residual silver paste, which was used during XPS sample preparation. Moreover, the high-resolution XPS spectrum of Sn 3d is plotted in Figure 2c, wherein two peaks at around 495.3 eV and 486.4 eV correspond to Sn $3d_{3/2}$ and Sn $3d_{5/2}$ [26,27], respectively, confirming the +4 valence state of Sn and formation of pure SnO_2 phase [23]. The XPS data were further verified by energy dispersive spectroscopy (EDS). The elemental distribution of Sn, O and Cu is shown in Figure 2d-f. It can be readily observed that the Sn and O signals are prominent in the regions of nanopillar, whereas the Cu signal is relatively weaker. Overall, these results confirm the formation of a phase pure SnO_2 nanoarray architecture.

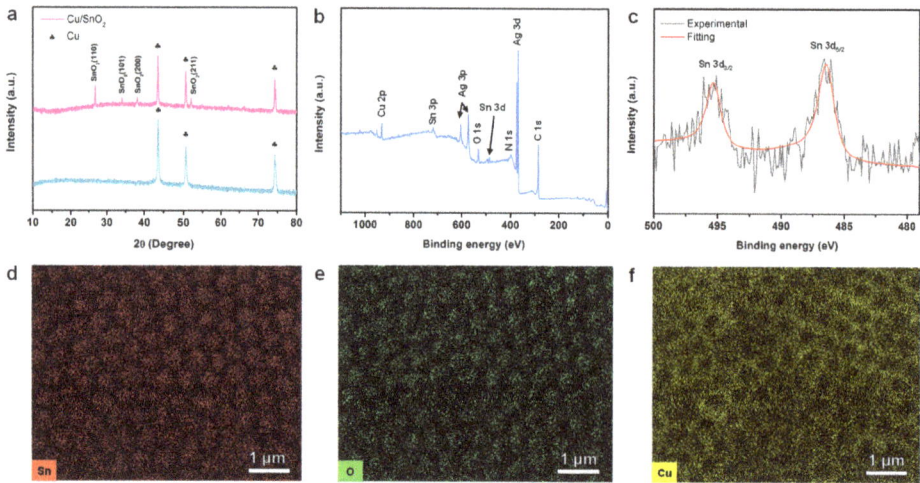

Figure 2. (a) XRD patterns of bare and SnO_2-coated Cu foils; (b) wide-range XPS spectrum of SnO_2 nanoarray; (c) high-resolution Sn 3d XPS spectrum; and elemental maps of (d) Sn, (e) O and (f) Cu from SnO_2 nanoarray.

After structural and microstructural characterization, the cyclic voltammetry (CV) and galvanostatic charge/discharge testing were carried out to evaluate the electrochemical performance of SnO_2 nanoarray anode in LIBs. Also, the electrochemical performance of SnO_2 thin film is studied to demonstrate the positive influence of SnO_2 nanoarray architecture. The first three CV curves were recorded at the scan rate of 0.2 mV s^{-1} in the voltage range of 0.01–3.0 V (vs. Li/Li$^+$) to investigate the redox process in SnO_2 nanoarray architecture. During the first cathodic scan, as shown in Figure 3a, two reduction peaks located at ~1.27 and 0.87 V (vs. Li/Li$^+$) correspond to the conversion reaction of SnO_2 to Sn (Equation (1)) and formation of solid electrolyte interphase (SEI), respectively [28,29]. Moreover, the peak at 0.18 V (vs. Li/Li$^+$) is attributed to the alloying process of Sn with Li to form Li$_x$Sn (Equation (2)) [30].

$$4Li^+ + SnO_2 + 4e^- \rightarrow 2Li_2O + Sn \qquad (1)$$

$$xLi^+ + Sn + xe^- \leftrightarrow Li_xSn \ (0 \leq x \leq 4.4) \qquad (2)$$

During the first anodic process, three oxidation peaks can be observed in Figure 3a, wherein the peak at ~0.51 V (vs. Li/Li$^+$) indicates the dealloying reaction of Li$_x$Sn ($0 \leq n \leq 4.4$) (Equation (2)) [31], the peak at ~1.25 V (vs. Li/Li$^+$) represents the decomposition of Li$_x$O, and the peak at ~1.91 V (vs. Li/Li$^+$) corresponds to the reoxidation of Sn [32]. The redox peaks in Figure 3a confirm the cathodic and anodic processes of the SnO_2 nanoarray, which are consistent with previously reported SnO_2-based anodes [20,25]. Moreover, the CV curves of SnO_2 thin film in Figure S1 (Supporting Information) show similar cathodic and anodic peaks, confirming the successful synthesis of phase pure

SnO$_2$ nanoarray. As shown in Figure 3a, the overlapping CV curves after the first cycle suggest the excellent reversibility of redox reactions during the charge/discharge process [11].

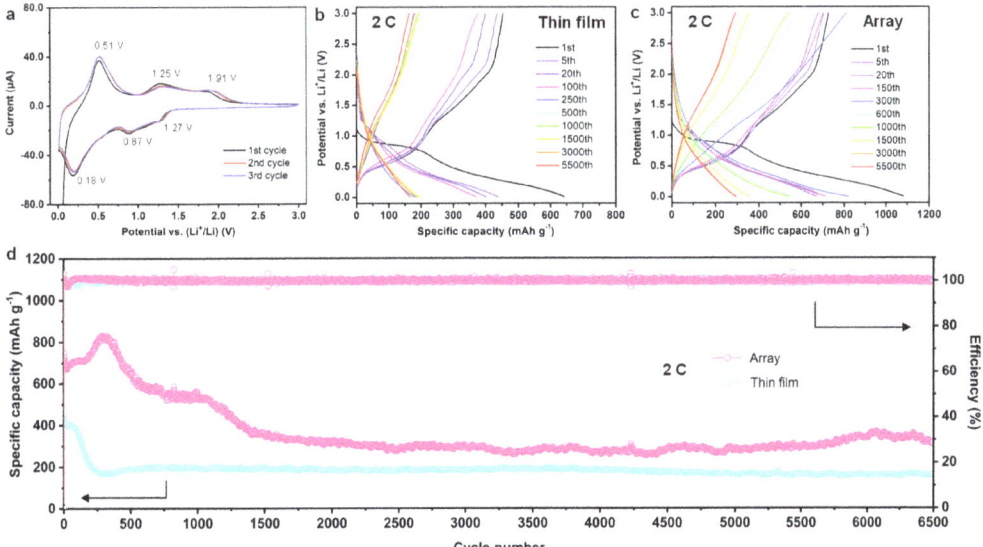

Figure 3. Electrochemical performance of SnO$_2$ electrodes. (**a**) CV curves of SnO$_2$ nanoarray at the scan rate of 0.2 mV s^{-1}; charge/discharge profiles of (**b**) PLD-fabricated SnO$_2$ thin film and (**c**) SnO$_2$ nanoarray anodes at the current density of 2 C; and (**d**) cyclic performance of PLD-fabricated SnO$_2$ thin film and SnO$_2$ nanoarray at the current density of 2 C.

Furthermore, the discharge/charge voltage profiles of SnO$_2$ thin film and SnO$_2$ nanoarray anode are plotted in Figure 3b,c, respectively. During the first discharge process, an apparent characteristic plateau at ~1.0 V (vs. Li/Li$^+$) is found (Figure 3b), which can be attributed to the irreversible occurrence of Equation (1) [25]. Moreover, a long slope due to the formation of Li$_x$Sn (Equation (2)) appeared with the further discharge process. The initial specific capacity of SnO$_2$ thin film anode reached ~642 mAh g^{-1}, which decreased to ~400 mAh g^{-1} and ~200 mAh g^{-1} after 100 and 250 cycles, respectively. The SnO$_2$ thin film anode rendered a stable specific capacity of ~200 mAh g^{-1} from 250 to 6500 charge/discharge cycles. These results suggest that the utilization of SnO$_2$ thin film as an anode in LIBs leads to excellent cyclic performance with a moderate specific capacity, which lays a solid foundation for the outstanding performance of SnO$_2$ nanoarray anode. Indeed, as shown in Figure 3c, the discharge/charge plateaus of SnO$_2$ nanoarray anode during the first cycle are located at the same positions as SnO$_2$ thin film. Moreover, the initial discharge capacity of SnO$_2$ nanoarray reached 1082 mAh g^{-1} and maintained at 700 mAh g^{-1}, 600 mAh g^{-1} and 300 mAh g^{-1} after 300, 600, and 1500 charge/discharge cycles, respectively, suggesting a remarkable improvement in electrochemical performance and cyclic stability of SnO$_2$ nanoarray. As presented in Figure S2, the superior electrochemical performance can be attributed to the unique architecture of SnO$_2$, where highly ordered SnO$_2$ nanopillar array provided adequate room for volumetric expansion and ensured structural integrity during the lithiation/delithiation process.

The excellent cyclic stability is further verified by carrying our charge/discharge cycling at the current density of 2 C (Figure 3d). In the case of SnO$_2$ thin film, the reversible specific capacity was first dropped from 642 mAh g^{-1} to 402 mAh g^{-1} after 20 discharge/charge cycles and then, the specific capacity was further dropped to about 180 mAh g^{-1} during 100–250 discharge/charge cycles, which remained stable until the

6500th charge/discharge cycle. These results suggest that the SnO_2 thin film can be used as an anode, which renders a stable but moderate specific capacity. Furthermore, the highly arrayed SnO_2 nanopillar delivered long cycling performance and improved reversible specific capacity. As shown in Figure 3d, under the same current density of 2 C, the initial specific capacity of SnO_2 nanoarray was found to be 1082 mAh g^{-1}, which dropped to 752 mAh g^{-1} during the 2nd charge/discharge cycle.

Interestingly, the specific capacity increased to 832 mAh g^{-1} at the 300th discharge/charge cycle after a slow decay, which is even higher than the 2nd reversible capacity. A possible reason for such an anomaly is that the thick SEI layer was exfoliated due to the high surface stress during the reactivation process. Subsequently, a fresh and thin SEI layer is formed [33], which can be partly confirmed from the detailed discussion of electrochemical impedance spectroscopy (Figure S3). The newly formed thin SEI layer is more stable than the previous thick SEI layer, which can bear severe volumetric change and fracture, resulting in improved Li storage capacity and longer cycling life [34]. Note that the stable charge-transfer resistance (R_{ct}) of SnO_2 in Figure S3 reveals the well-maintained interface between SnO_2 nanoarray and substrate. Indeed, as shown in Figure 3d, the specific capacity of SnO_2 nanoarray gradually decreased and became stable at 524 mAh g^{-1} during the 300th to 1100th charge/discharge cycles. The attained specific capacity is two times higher than the SnO_2 thin film. With further cycling, the reversible specific capacity stabilized at ~313 mAh g^{-1} from 2000th to 6500th cycles, demonstrating the long cycling life and high capacity of binder-free SnO_2 nanoarray anode.

A range of current densities, i.e., 0.5 C to 16 C, was selected to investigate the cyclability of SnO_2 nanoarray and SnO_2 thin film anodes. Overall, the SnO_2 thin film delivered a lower discharge capacity than SnO_2 nanoarray, as shown in Figure 4a. In the case of SnO_2 thin film, the discharge specific capacity decreased with increasing current density from 440 mAh g^{-1} at 0.5 C to 314 mAh g^{-1} at 16 C. The SnO_2 thin film could not recover the initial capacity when the current density was reduced after high-rate cycling, which can be ascribed to structural failure of SnO_2 thin film. On the other side, the SnO_2 nanoarray anode rendered a high discharge capacity of 536 mAh g^{-1} at 0.5 C and 414 mAh g^{-1} at 16 C. More importantly, the SnO_2 nanoarray maintained a reversible capacity of 520 mAh g^{-1} and 517 mAh g^{-1}, under 1 C, after 100th and 160th cycles after high rate cycling at 16 C, indicating the excellent rate capability of nanoarray architecture. The difference in rate performance of the SnO_2 thin film and SnO_2 nanoarray can be attributed to the architectural differences, where the highly arrayed SnO_2 nanopillar enjoy enough space to alleviate volumetric expansion and ensure structural stability during the charge/discharge process.

Furthermore, we studied the rate capability of SnO_2 nanoarray at extremely high current densities, ranging from 2 C to 64 C. As shown in Figure 4b, the SnO_2 nanoarray delivered a specific capacity of 530, 471, 442, 398, and 354 mAh g^{-1} at the current density of 2 C, 4 C, 8 C, 16 C, and 32 C during the 1st to 50th cycle, as well as a reversible specific capacity of 556, 498, 448, 404 and 354 mAh g^{-1} at 2 C, 4 C, 8 C, 16 C and 32 C during 50th and 100th cycle, suggesting the high-rate capability of SnO_2 nanoarray. Moreover, at an ultrahigh current density of 64 C, the SnO_2 nanoarray delivered a specific capacity of ~278 mAh g^{-1} during the 140th and 150th cycle (Figure 4b) and then, recovered a high specific capacity of 670 mAh g^{-1} at the current density of 2 C (200th cycle). These results confirm that the SnO_2 nanoarray anode can work under ultrahigh current densities and possess remarkable rate capability. Lastly, the electrochemical performance of binder-free SnO_2 anode is compared with previously reported SnO_2-based anodes [14,24,31,35–37], as shown in Figure 4c, demonstrating the superior Li storage capacity and high-rate capability of the as-fabricated SnO_2 nanoarray. The superior electrochemical performance of SnO_2 nanoarray can be attributed to the unique architecture of SnO_2, where highly ordered SnO_2 nanopillar array provided adequate room for volumetric expansion and ensured structural integrity during the lithiation/delithiation process. It also should note that the other two points may also play important role for our SnO_2 nanoarray to resist the 300% volumetric change during charge/discharge process. First, we fabricate SnO_2 nanopillar based on

PLD can ensure excellent contact between SnO$_2$ and Cu, but it will also bring a strong substrate constraint from the Cu substrate to SnO$_2$, which will greatly clamp the volume expansion of SnO$_2$ along lateral direction, making the volumetric change less than 300%. Moreover, the obtained SnO$_2$ nanopillars usually are not in uniform cylindrical shape when we use AAO-assisted PLD, as we schematic shown in Figure 1a, the top parts (away from interface of SnO$_2$ and Cu) of SnO$_2$ nanopillars usually prefer to appear in pyramid shape, such unique architecture provide more room and multiple routes for the expansion of SnO$_2$ nanopillars.

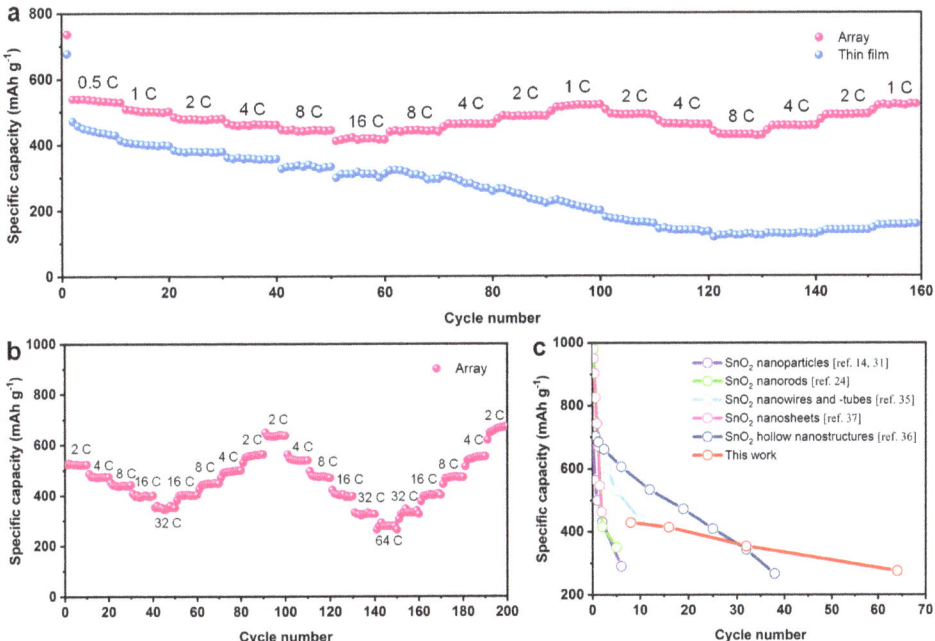

Figure 4. Rate performance of SnO$_2$ nanoarray and SnO$_2$ thin films. (**a**) The rate capability of SnO$_2$ nanoarray and PLD-fabricated SnO$_2$ thin film at different current densities, ranging from 0.5 C to 16 C; (**b**) rate capability of SnO$_2$ nanoarray at different current densities, ranging from 2 C to 64 C; and (**c**) comparison of the electrochemical performance for SnO$_2$-based anodes.

4. Conclusions

In summary, we have fabricated a well-oriented and binder-free SnO$_2$ nanopillar array as an anode electrode for LIBs. The as-synthesized SnO$_2$ nanoarray anode rendered superior Li storage capacity of 1082 mAh g^{-1} and excellent cyclic stability (~313 mAh g^{-1} after 6500 charge/discharge cycles). Moreover, the SnO$_2$ nanoarray demonstrated ultrahigh rate capability, i.e., the capacity of ~278 mAh g^{-1} at the current density of 64 C, outperforming the previously reported binder-free SnO$_2$-based anodes. The superior electrochemical performance of SnO$_2$ nanoarray can be attributed to the unique architecture of SnO$_2$, where the highly arrayed SnO$_2$ nanopillar provided adequate room for volumetric expansion and ensured structural integrity during the lithiation/delithiation process. The fabrication of SnO$_2$ nanoarray presents an effective approach to enhance the energy density and rapid charge/discharge capability of LIBs.

Supplementary Materials: The following are available online at https://www.mdpi.com/article/10.3390/nano11051307/s1, Figure S1: Cyclic voltammetry curves of SnO$_2$ thin films at scan rate 0.2 mV s^{-1}, Figure S2: Top-view SEM images. (a,b) the top-view SEM images of as-deposited (a)

SnO$_2$ thin film and (b) after 5 discharge/charge cycles; (c,d) the top-view SEM images of as-deposited (c) SnO$_2$ nanopillar array and (d) after 5 discharge/charge cycles, Figure S3: (a,b) the nyquist plots of SnO$_2$ nanoarray (a) and thin film (b) at 20th, 50th, 400th cycles.

Author Contributions: L.D.: Methodology, Investigation, Resources, Data analysis, Writing-original draft, Writing-review & editing. X.Z.: Investigation, Supervision. J.Z.: Data analysis, Investigation. B.F.: Data analysis, Investigation. Y.S.: Data analysis, Investigation. C.R.: Data analysis, Investigation. J.W.: Investigation, Supervision. G.Z.: Conceptualization, Funding acquisition, Project administration, Supervision, Writing-original draft, Writing-review & editing. All authors have read and agreed to the published version of the manuscript.

Funding: This work was supported by the National Natural Science Foundation of China (No. 51902337, No. 92066102, No. 11875229), Shenzhen Science and Technology Innovation Committee (Nos. KQTD20170810160424889, Nos. RCYX20200714114733204), Shenzhen Institute of Advanced Technology, Chinese Academy of Sciences (No. Y9G050).

Data Availability Statement: The data presented in this study are available on request from the corresponding author.

Conflicts of Interest: The authors declare that they have no known competing financial interests or personal relationships that could have appeared to influence the work reported in this paper.

References

1. Zhang, D.; Xu, X.; Qin, Y.; Ji, S.; Huo, Y.; Wang, Z.; Shen, J.; Liu, J. Recent progress in organic-inorganic composite solid electrolytes for all-solid-state lithium batteries. *Chem. Eur. J.* **2020**, *26*, 1720–1736. [CrossRef] [PubMed]
2. Tarascon, J.-M.; Recham, N.; Armand, M.; Chotard, J.-N.; Barpanda, P.; Walker, W.; Dupont, L. Hunting for Better Li-Based Electrode Materials via Low Temperature Inorganic Synthesis. *Chem. Mater.* **2010**, *22*, 724–739. [CrossRef]
3. Sui, Y.; Liu, C.; Masse, R.C.; Neale, Z.G.; Atif, M.; AlSalhi, M.; Cao, G. Dual-ion batteries: The emerging alternative rechargeable batteries. *Energy Storage Mater.* **2020**, *25*, 1–32. [CrossRef]
4. Huang, X.; Yu, H.; Chen, J.; Lu, Z.; Yazami, R.; Hng, H.H. Ultrahigh Rate Capabilities of Lithium-Ion Batteries from 3D Ordered Hierarchically Porous Electrodes with Entrapped Active Nanoparticles Configuration. *Adv. Mater.* **2014**, *26*, 1296–1303. [CrossRef]
5. Jiang, Y.; Hu, M.; Zhang, D.; Yuan, T.; Sun, W.; Xu, B.; Yan, M. Transition metal oxides for high performance sodium ion battery anodes. *Nano Energy* **2014**, *5*, 60–66. [CrossRef]
6. Zheng, M.; Tang, H.; Li, L.; Hu, Q.; Zhang, L.; Xue, H.; Pang, H. Hierarchically Nanostructured Transition Metal Oxides for Lithium-Ion Batteries. *Adv. Sci.* **2018**, *5*, 1700592. [CrossRef]
7. Zhong, G.; Qu, K.; Ren, C.; Su, Y.; Fu, B.; Zi, M.; Dai, L.; Xiao, Q.; Xu, J.; Zhong, X.; et al. Epitaxial array of Fe$_3$O$_4$ nanodots for high rate high capacity conversion type lithium ion batteries electrode with long cycling life. *Nano Energy* **2020**, *74*, 104876. [CrossRef]
8. Wu, Z.S.; Ren, W.; Wen, L.; Gao, L.; Zhao, J.; Chen, Z.; Zhou, G.; Li, F.; Cheng, H.M. Graphene anchored with Co$_3$O$_4$ nanoparticles as anode of lithium ion batteries with enhanced reversible capacity and cyclic performance. *ACS Nano* **2010**, *4*, 3187–3194. [CrossRef]
9. Hu, R.; Ouyang, Y.; Liang, T.; Wang, H.; Liu, J.; Chen, J.; Zhu, M. Stabilizing the nanostructure of SnO$_2$ anodes by transition metals: A route to achieve high initial coulombic efficiency and stable capacities for lithium storage. *Adv. Mater.* **2017**, *29*, 1605006. [CrossRef]
10. Courtney, I.A.; Dahn, J.R. Electrochemical and in situ X-ray diffraction studies of the reaction of lithium with tin oxide com-posites. *J. Electrochem. Soc.* **1997**, *144*, 2045–2052. [CrossRef]
11. Zhou, X.; Wan, L.-J.; Guo, Y.-G. Binding SnO$_2$ Nanocrystals in Nitrogen-Doped Graphene Sheets as Anode Materials for Lithium-Ion Batteries. *Adv. Mater.* **2013**, *25*, 2152–2157. [CrossRef] [PubMed]
12. Idota, Y.; Kubota, T.; Matsufuji, A.; Maekawa, Y.; Miyasaka, T. Tin-Based Amorphous Oxide: A High-Capacity Lithium-Ion-Storage Material. *Science* **1997**, *276*, 1395–1397. [CrossRef]
13. Ng, S.; dos Santos, D.; Chew, V.; Wexler, D.; Wang, J.; Dou, S.; Liu, H. Polyol-mediated synthesis of ultrafine tin oxide nanoparticles for reversible Li-ion storage. *Electrochem. Commun.* **2007**, *9*, 915–919. [CrossRef]
14. Etacheri, V.; Seisenbaeva, G.A.; Caruthers, J.; Daniel, G.; Nedelec, J.-M.; Kessler, V.G.; Pol, V.G. Ordered Network of Interconnected SnO$_2$ Nanoparticles for Excellent Lithium-Ion Storage. *Adv. Energy Mater.* **2015**, *5*, 1401289. [CrossRef]
15. Wang, H.-G.; Wu, Q.; Wang, Y.; Wang, X.; Wu, L.; Song, S.; Zhang, H. Molecular Engineering of Monodisperse SnO$_2$ Nanocrystals Anchored on Doped Graphene with High-Performance Lithium/Sodium-Storage Properties in Half/Full Cells. *Adv. Energy Mater.* **2019**, *9*, 1802993. [CrossRef]
16. Zhong, G.; Bitla, Y.; Wang, J.; Zhong, X.; An, F.; Chin, Y.-Y.; Zhang, Y.; Gao, W.; Zhang, Y.; Eshghinejad, A.; et al. Tuning Fe concentration in epitaxial gallium ferrite thin films for room temperature multiferroic properties. *Acta Mater.* **2018**, *145*, 488–495. [CrossRef]
17. Li, Z.; Wang, Y.; Tian, G.; Li, P.; Zhao, L.; Zhang, F.; Yao, J.; Fan, H.; Song, X.; Chen, D.; et al. High-density array of ferroelectric nanodots with robust and reversibly switchable topological domain states. *Sci. Adv.* **2017**, *3*, e1700919. [CrossRef]

18. Zhong, G.; An, F.; Bitla, Y.; Wang, J.; Zhong, X.; Ye, M.; Zhang, Y.; Gao, W.; Pan, X.; Xie, S.; et al. Self-assembling epitaxial growth of a single crystalline $CoFe_2O_4$ nanopillar array via dual-target pulsed laser deposition. *J. Mater. Chem. C* **2018**, *6*, 4854–4860. [CrossRef]
19. Chen, S.; Wang, M.; Ye, J.; Cai, J.; Ma, Y.; Zhou, H.; Qi, L. Kinetics-controlled growth of aligned mesocrystalline SnO_2 nanorod arrays for lithium-ion batteries with superior rate performance. *Nano Res.* **2013**, *6*, 243–252. [CrossRef]
20. Park, M.-S.; Wang, G.-X.; Kang, Y.-M.; Wexler, D.; Dou, S.-X.; Liu, H.-K. Preparation and Electrochemical Properties of SnO_2 Nanowires for Application in Lithium-Ion Batteries. *Angew. Chem. Int. Ed.* **2007**, *46*, 750–753. [CrossRef]
21. Ying, Z.; Wan, Q.; Cao, H.; Song, Z.T.; Feng, S.L. Characterization of SnO_2 nanowires as an anode material for Li-ion batteries. *Appl. Phys. Lett.* **2005**, *87*, 113108. [CrossRef]
22. Wang, Y.; Lee, J.Y.; Zeng, H.C. Polycrystalline SnO_2 Nanotubes Prepared via Infiltration Casting of Nanocrystallites and Their Electrochemical Application. *Chem. Mater.* **2005**, *17*, 3899–3903. [CrossRef]
23. Hwang, S.M.; Lim, Y.-G.; Kim, J.-G.; Heo, Y.-U.; Lim, J.H.; Yamauchi, Y.; Park, M.-S.; Kim, Y.-J.; Dou, S.X.; Kim, J.H. A case study on fibrous porous SnO_2 anode for robust, high-capacity lithium-ion batteries. *Nano Energy* **2014**, *10*, 53–62. [CrossRef]
24. Liu, J.; Li, Y.; Huang, X.; Ding, R.; Hu, Y.; Jiang, J.; Liao, L. Direct growth of SnO_2 nanorod array electrodes for lithium-ion batteries. *J. Mater. Chem.* **2009**, *19*, 1859–1864. [CrossRef]
25. Park, M.-S.; Kang, Y.-M.; Wang, G.-X.; Dou, S.-X.; Liu, H.-K. The Effect of Morphological Modification on the Electrochemical Properties of SnO_2 Nanomaterials. *Adv. Funct. Mater.* **2008**, *18*, 455–461. [CrossRef]
26. Liu, X.; Jiang, Y.; Li, K.; Xu, F.; Zhang, P.; Ding, Y. Electrospun free-standing N-doped C@SnO_2 anode paper for flexible Li-ion batteries. *Mater. Res. Bull.* **2019**, *109*, 41–48. [CrossRef]
27. Aravindan, V.; Jinesh, K.; Prabhakar, R.R.; Kale, V.S.; Madhavi, S. Atomic layer deposited (ALD) SnO_2 anodes with exceptional cycleability for Li-ion batteries. *Nano Energy* **2013**, *2*, 720–725. [CrossRef]
28. Li, N.; Martin, C.R. A High-Rate, High-Capacity, Nanostructured Sn-Based Anode Prepared Using Sol-Gel Template Synthesis. *J. Electrochem. Soc.* **2001**, *148*, A164–A170. [CrossRef]
29. Shi, W.; Lu, B. Nanoscale Kirkendall effect synthesis of echinus-like SnO_2@SnS_2 nanospheres as high-performance anode ma-terial for lithium ion batteries. *Electrochim. Acta* **2014**, *133*, 247–253. [CrossRef]
30. Lin, J.; Peng, Z.; Xiang, C.; Ruan, G.; Yan, Z.; Natelson, D.; Tour, J.M. Graphene Nanoribbon and Nanostructured SnO_2 Composite Anodes for Lithium Ion Batteries. *ACS Nano* **2013**, *7*, 6001–6006. [CrossRef]
31. Yin, L.; Chai, S.; Wang, F.; Huang, J.; Li, J.; Liu, C.; Kong, X. Ultrafine SnO_2 nanoparticles as a high performance anode material for lithium ion battery. *Ceram. Int.* **2016**, *42*, 9433–9437. [CrossRef]
32. Ferraresi, G.; Villevieille, C.; Czekaj, I.; Horisberger, M.; Novák, P.; El Kazzi, M. SnO_2 Model Electrode Cycled in Li-Ion Battery Reveals the Formation of Li_2SnO_3 and Li_8SnO_6 Phases through Conversion Reactions. *ACS Appl. Mater. Interfaces* **2018**, *10*, 8712–8720. [CrossRef]
33. Nadimpalli, S.P.V.; Sethuraman, V.A.; Bucci, G.; Srinivasan, V.; Bower, A.F.; Guduru, P.R. On Plastic Deformation and Fracture in Si Films during Electrochemical Lithiation/Delithiation Cycling. *J. Electrochem. Soc.* **2013**, *160*, A1885–A1893. [CrossRef]
34. Sun, H.; Xin, G.; Hu, T.; Yu, M.; Shao, D.; Sun, X.; Lian, J. High-rate lithiation-induced reactivation of mesoporous hollow spheres for long-lived lithium-ion batteries. *Nat. Commun.* **2014**, *5*, 4526. [CrossRef]
35. Ko, Y.-D.; Kang, J.-G.; Park, J.-G.; Lee, S.; Kim, D.-W. Self-supported SnO_2 nanowire electrodes for high-power lithium-ion batteries. *Nanotechnology* **2009**, *20*, 455701. [CrossRef] [PubMed]
36. Park, G.D.; Lee, J.-K.; Kang, Y.C. Synthesis of Uniquely Structured SnO_2 Hollow Nanoplates and Their Electrochemical Properties for Li-Ion Storage. *Adv. Funct. Mater.* **2016**, *27*, 1603399. [CrossRef]
37. Zhu, Y.; Guo, H.; Zhai, H.; Cao, C. Microwave-Assisted and Gram-Scale Synthesis of Ultrathin SnO_2 Nanosheets with Enhanced Lithium Storage Properties. *ACS Appl. Mater. Interfaces* **2015**, *7*, 2745–2753. [CrossRef] [PubMed]

Article

Impact of Surface Chemistry of Silicon Nanoparticles on the Structural and Electrochemical Properties of Si/Ni$_{3.4}$Sn$_4$ Composite Anode for Li-Ion Batteries

Tahar Azib [1], Claire Thaury [1,2], Fermin Cuevas [1,*], Eric Leroy [1], Christian Jordy [2], Nicolas Marx [3] and Michel Latroche [1]

1. Univ Paris Est Creteil, CNRS, ICMPE, UMR 7182, 2 rue Henri Dunant, 94320 Thiais, France; azib@icmpe.cnrs.fr (T.A.); claire.thaury@gmail.com (C.T.); leroy@icmpe.cnrs.fr (E.L.); latroche@icmpe.cnrs.fr (M.L.)
2. SAFT Batteries, 113 Bd. Alfred Daney, 33074 Bordeaux, France; Christian.JORDY@saftbatteries.com
3. Umicore, Watertorenstraat 33, 2250 Olen, Belgium; Nicolas.Marx@eu.umicore.com
* Correspondence: cuevas@icmpe.cnrs.fr

Abstract: Embedding silicon nanoparticles in an intermetallic matrix is a promising strategy to produce remarkable bulk anode materials for lithium-ion (Li-ion) batteries with low potential, high electrochemical capacity and good cycling stability. These composite materials can be synthetized at a large scale using mechanical milling. However, for Si-Ni$_3$Sn$_4$ composites, milling also induces a chemical reaction between the two components leading to the formation of free Sn and NiSi$_2$, which is detrimental to the performance of the electrode. To prevent this reaction, a modification of the surface chemistry of the silicon has been undertaken. Si nanoparticles coated with a surface layer of either carbon or oxide were used instead of pure silicon. The influence of the coating on the composition, (micro)structure and electrochemical properties of Si-Ni$_3$Sn$_4$ composites is studied and compared with that of pure Si. Si coating strongly reduces the reaction between Si and Ni$_3$Sn$_4$ during milling. Moreover, contrary to pure silicon, Si-coated composites have a plate-like morphology in which the surface-modified silicon particles are surrounded by a nanostructured, Ni$_3$Sn$_4$-based matrix leading to smooth potential profiles during electrochemical cycling. The chemical homogeneity of the matrix is more uniform for carbon-coated than for oxygen-coated silicon. As a consequence, different electrochemical behaviors are obtained depending on the surface chemistry, with better lithiation properties for the carbon-covered silicon able to deliver over 500 mAh/g for at least 400 cycles.

Keywords: Li-ion batteries; anodes; intermetallics; silicon; composites; nanomaterials; coating; mechanochemistry

1. Introduction

The rapid development of portable electronics, Electric Vehicles (EVs) and renewable energies requires light, safe and high-capacity rechargeable energy storage devices such as lithium-ion (Li-ion) batteries, one of the most efficient electrochemical storage systems today [1]. However, Li-ion batteries still need to be improved regarding design, electrode capacities and electrolyte stability [2]. Carbon-based anode materials are cheap and easy to prepare but suffer from moderate capacity (372 mAhg^{-1} for graphite), which remains a limitation for the development of high-energy density storage [3]. Moreover, graphite suffers from parasitic reaction with the liquid electrolyte during charging and discharging processes to form the so-called Solid Electrolyte Interface (SEI), growth of which is detrimental for the stability and the capacity of the battery [4]. Therefore, new anode materials are required for the development of high-capacity Li-ion batteries.

Three main types of anode materials are currently envisaged for the replacement of graphite. Firstly, there are novel carbonaceous-based materials such as carbon nanotubes, carbon nanospheres, graphene and porous graphitic carbon [5–8]. Secondly, there are conversion-type transition metal compounds such as transition metal oxides, sulphides, selenides, fluorides, nitrides, phosphides and hydrides [9,10], and finally, there are silicon and tin-based anodes [11,12].

Pure *p*-type elements like Si or Sn are considered as promising to develop negative electrodes for Li-ion batteries [12–14]. Indeed, they can both be lithiated [15] to form binary compounds ($Li_{4.4}Sn$ and $Li_{3.75}Si$) with very large electrochemical capacities (994 and 3600 mAhg^{-1}, respectively) [12,16,17]. In addition to their high theoretical capacity, these elements have low potential and environmental friendliness. However, Si electrodes suffer from severe volume expansion during lithiation (up to 400%) [18]. Such swelling induces several drawbacks from the very first cycles like amorphization, delamination and capacity degradation, which are unfavorable for long term cycling [19–21].

To overcome these drawbacks, embedding the capacitive elements in a metallic matrix able to provide good electronic conductivity and to hold the volume changes is a beneficial solution [22]. This can be done with binary compounds having one element reacting with Li when the other one remains inactive, like for Ni_3Sn_4 [23–25], Cu_6Sn_5 [26], $CoSn_2$ [27], $FeSn_2$ [12], $NiSi_2$ [28] or $TiSi_2$ [29].

Following this concept, our group thoroughly investigated composites of general formulation Si-$Ni_{3.4}Sn_4$-Al-C prepared via mechanochemistry [30–32]. They consist of submicronic silicon particles embedded in a nanostructured matrix made of $Ni_{3.4}Sn_4$, aluminum and graphite carbon. As reported by [33], low aluminum content (~3 wt.%) improves the cycle life of Si-Sn-type anodes. Carbon addition acts as a Process Control Agent (PCA), minimizing reactivity between Si and $Ni_{3.4}Sn_4$ on milling [30]. These composites take advantage of the high capacities of silicon and tin, the good ionic and electronic conductivity of the matrix and its elastic properties to manage volume expansion. Further improvement for these composites can be foreseen by playing with the surface chemistry of the silicon particles [34,35].

In the present work, we investigate an alternative approach to PCA addition on milling that consists of modifying the surface chemistry of the silicon particles used for the composite synthesis. The Si surface is covered either with a carbonaceous or an oxide layer. Structural, morphological and electrochemical properties of these surface-modified silicon composites have been fully characterized and compared to those of non-modified Si. These new composites have very different properties, giving the best electrochemical performances for the carbon-coated silicon.

2. Materials and Methods

Three composites of Si-$Ni_{3.4}Sn_4$-Al were prepared using mechanochemistry of intermetallic $Ni_{3.4}Sn_4$ (75 wt.%), Al (3 wt.%) and three different kinds of silicon (22 wt.%): bare silicon, carbon- and oxide-coated silicon. Bare Si was provided by SAFT (purity 99.9%) as a reference for this work and is hereafter labelled as Si_R. The second Si precursor, labelled as Si_C, was provided by Umicore. Silicon particles were coated with carbon via Chemical Vapor Deposition (CVD) at 800 °C for 3 h. The third one (Si_O) was purchased from MTI Corporation (CA, USA) as pure silicon. However, chemical analysis revealed that the particles were covered by a thin oxide layer. They were thus fully characterized regarding their surface chemistry and used as a Si-surface oxidized precursor (Si_O). These three Si-precursors were used to synthetize the composites (Si-$Ni_{3.4}Sn_4$-Al) via ball milling of Si, $Ni_{3.4}Sn_4$ (99.9%, ≤125 µm, home-made) and Al (99%, ≤75 µm, Sigma-Aldrich, Saint-Louis, USA) powders for 20 h under an inert atmosphere. Further details on intermetallic $Ni_{3.4}Sn_4$ and composite synthesis can be found in [30–32]. No addition of carbon graphite in the milling jar as PCA was used for the current investigation. The obtained composites are labelled as Si_R-NiSn, Si_C-NiSn and Si_O-NiSn, respectively.

X-Ray Diffraction (XRD) analysis of Si powders and composite materials was done with a Bruker D8 θ-θ diffractometer (Karlsruhe, Germany) using Cu-Kα radiation, in a 2θ range from 20 to 100° with a step size of 0.02°. Diffraction patterns were analyzed using the Rietveld method using the FULLPROF package [36]. Morphology of the composites was studied using Scanning Electron Microscopy (SEM) using a SEM-FEG MERLIN from Zeiss (Jena, Germany). Images were acquired from either Secondary Electrons (SE) or Back-Scattered Electrons (BSE) to provide information on particle morphology as well as phase distribution. Microstructural and chemical properties were analyzed using Transmission Electron Microscopy (TEM) with a Tecnai FEI F20 ST microscope (Hillsboro, OR, USA) providing high spatial resolution imaging of the scale morphology as well as chemistry via Energy-Dispersive X-ray spectroscopy (EDX) analyses. Images were taken in both bright and dark fields. Elemental mapping analysis was carried out using EDX analysis in Scanning Transmission Electron Microscopy (STEM) mode and via Electron Energy Filtered Transmission Electron Microscopy (EFTEM). The samples were prepared by mixing the composite with Cu powder, followed by cold-rolling and thinning with argon ions in a GATAN precision ion polishing system.

Electrochemical measurements were carried out via galvanostatic cycling in half-coin type cells. A working electrode was prepared by mixing 40 wt.% of the 20-h-milled composite sieved under 36 µm, 30 wt.% of carboxymethyl-cellulose (CMC) binder and 30 wt.% of carbon black. Low loading of active material was adopted to avoid limitations on electrochemical performance due to electrode formulation. Metallic lithium was used as counter negative electrode separated by a 1 M solution electrolyte of $LiPF_6$ dissolved in Ethylene Carbonate (EC)/Propylene Carbonate (PC)/Dimethyl Carbonate (DMC) (1:1:3 $v/v/v$), supported by a microporous polyolefin Celgard™ membrane and a nonwoven polyolefin separator. The EC/PC/DMC mixture of carbonate-based solvent was selected based on its outstanding physico-chemical properties [37]. The battery was assembled in an argon filled glove box. The experiments were performed using a Biologic (Seyssinet-Pariset, France) potentiostat instrument. To ensure full electrode lithiation, cells were cycled at C/50 for the first cycle, with a voltage window comprised between 0 and 2 V vs. Li^+/Li, and at C/20 for the second and third cycles, with a voltage window comprised between 70 mV and 2 V vs. Li^+/Li. The cut-off voltage of 70 mV was imposed to avoid the formation of crystalline $Li_{15}Si_4$ phase [38]. For all subsequent cycles, the kinetic regime was increased to C/10 to accomplish long-term cycling studies (up to 400 cycles) in a reasonable time duration and with a voltage window comprised between 70 mV and 2 V. Reference cycles at a rate of C/20 were done at second and third cycles and after every 20 cycles. Only the first and reference cycles are reported in this paper.

3. Results
3.1. Chemical and Microstructural Characterization
3.1.1. Characterization of Bare and Surface-Modified Si Nanopowders

The XRD pattern of the Si_R sample is displayed in Figure S1a (Supplementary Materials (SM)). All peaks can be indexed in the cubic space group Fd-$3m$ with lattice constant $a = 5.426 \pm 2$ Å, slightly smaller than the well-crystallized silicon standard ($a = 5.430$ Å [39]). The measured crystallite size deduced from the diffraction peak linewidths is 16 ± 2 nm. SEM images reveal that the Si_R powder has an interconnected worm-like morphology (Figure S1b). When observed using TEM, round particles with an average size of 180 nm are observed (Figure S1c). EFTEM analysis shows pure Si material with minor traces of oxygen at the surface (Figure S1d).

The Si_C particles were chemically analyzed and contained Si (68 wt.%), C (30 wt.%) and O (2 wt.%). The Rietveld analysis of XRD patterns of Si_C is shown in Figure S2a. The main phase is silicon ($a = 5.429 \pm 2$ Å, space group Fd-$3m$). The crystallite size is 79 ± 3 nm. A small and broad peak around 25°-2θ is attributed to the presence of poorly crystallized graphite (Figure S2a). SEM analysis reveals that the Si_C powder is made of large agglomerates (10 to 100 µm) of primary spherical particles (Figure S2b). The particles were further investigated via TEM, confirming their spherical morphology (Figure S2c). In addition, TEM images show that the particles have an average size of 50 nm and are covered by a thin layer measuring a few nanometers (~10 nm) (Figure S2d).

Elemental mapping indicates that the core of the particle is made of silicon surrounded by a thin carbon shell. At the interface, a composition gradient exists revealing a possible formation of silicon carbide SiC, assuring the chemical bonding between the two elemental layers.

The Si_O particles were analyzed using XRD, showing that the main phase is silicon (cubic phase; $a = 5.426 \pm 2$ Å; space group Fd-$3m$) with a crystallite size around 26 ± 1 nm (Figure S3a). SEM analysis shows that the powder is made of large agglomerates up to 50 µm formed by primary submicrometric rounded particles (Figure S3b). From TEM analysis, it is observed that the primary particles are spherical with an average size of 70 nm (Figure S3c). Elemental analysis indicates a core of silicon surrounded by a shell containing both oxygen and silicon, which is attributed to the formation of SiO_2 (Figure S3d). A rough estimation based of the relative sizes of the core (58 nm in diameter) and the shell (6 nm thick) as well as the crystal densities of Si (2.33 g/cm^3) and SiO_2 (2.65 g/cm^3) leads to a global oxygen content in the Si_O sample of 25 wt.%.

To summarize, Figure 1 displays TEM elemental mapping for the three types of Si nanopowders used in this study: bare Si, showing minor traces of oxygen at the surface, carbon-coated Si with a 10-nm-thick carbonaceous layer and oxide-coated Si with a 6-nm-thick SiO_2 oxide shell.

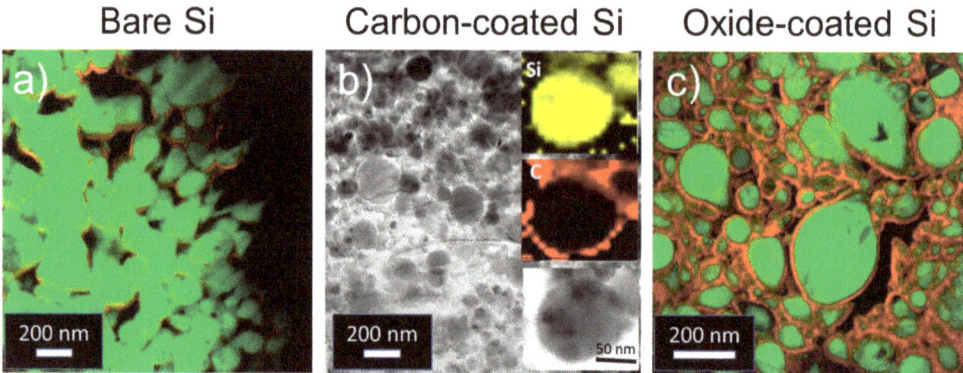

Figure 1. Transmission Electron Microscopy (TEM) images and elemental mapping of Si nanoparticles used as precursors for composite synthesis. (**a**) bare Si, Si_R (Si in green, oxygen traces in red), (**b**) carbon-coated Si, Si_C (Si in yellow, carbon in red) and (**c**) oxide-coated Si, Si_O (Si in green, oxygen in red).

3.1.2. Characterization of the Composite Materials

Three composites, Si_R-NiSn, Si_C-NiSn and Si_O-NiSn, were synthetized using the previously analyzed silicon nanopowders. The weight and atomic compositions of the composites are given in Table 1. The amount of carbon and oxygen for each composite are estimated from the chemical analysis of the surface-modified Si particles assuming a shell of pure C for Si_C and a shell of SiO_2 for Si_O.

Table 1. Compositions of Si_R-NiSn, Si_C-NiSn and Si_O-NiSn composites.

Composites	Weight Composition	Atomic Composition
Si_R-NiSn	$Si_{0.22}Ni_{0.22}Sn_{0.53}Al_{0.03}$	$Si_{0.46}Ni_{0.22}Sn_{0.26}Al_{0.06}$
Si_C-NiSn	$Si_{0.15}Ni_{0.22}Sn_{0.53}Al_{0.03}C_{0.07}$	$Si_{0.26}Ni_{0.18}Sn_{0.22}Al_{0.06}C_{0.28}$
Si_O-NiSn	$Si_{0.17}Ni_{0.22}Sn_{0.53}Al_{0.03}O_{0.06}$	$Si_{0.32}Ni_{0.20}Sn_{0.24}Al_{0.05}O_{0.19}$

Evolution of the diffractograms for the three composites as a function of milling time between 1 and 20 h is shown in Figure S4. For Si_R-NiSn, diffraction peaks of the intermetallic precursor $Ni_{3.4}Sn_4$ progressively disappear, Si peaks broaden and new peaks due to Sn formation appear. For Si$_C$-NiSn, diffraction peaks of Ni$_{3.4}$Sn$_4$ and Si are preserved though undergoing significant line broadening. A minor contribution of Sn formation is detected. Finally, for Si$_O$-NiSn, Ni$_{3.4}$Sn$_4$ and Si are mostly preserved, but compared to the Si$_C$-NiSn composite, peak broadening for the intermetallic precursor is less pronounced and a secondary intermetallic phase of Ni$_3$Sn$_4$ with lower Ni-content than that of the pristine precursor Ni$_{3.4}$Sn$_4$ is formed. A minor contribution of Sn formation is also detected.

The XRD diffraction patterns for the 20-h-milled composites are displayed in Figure 2. Rietveld analysis is provided in Figure S5 and collected crystal data are gathered in Table 2. For the composite made with bare Si, Si$_R$-NiSn, major phases are Sn (43 ± 1 wt.%) and NiSi$_2$ (35 ± 1 wt.%). These phases result from a mechanically-induced chemical reaction between the milling precursors Ni$_{3.4}$Sn$_4$ and Si. In contrast, for the coated composites Si$_C$-NiSn and Si$_O$-NiSn, the main phase remains Ni$_{3+x}$Sn$_4$-type (~85 wt.%) evidencing minor chemical reaction between Ni$_{3.4}$Sn$_4$ and Si on milling. Indeed, after 20 h of milling, the content of Sn byproduct in the Si-coated composites is as low as ~3 wt.%. Nonetheless, it is worth noticing that for the Si$_O$-NiSn composite, almost half of the pristine intermetallic precursor Ni$_{3.4}$Sn$_4$ (34 ± 2 wt.%) diminishes in Ni-content to form Ni$_3$Sn$_4$. In addition, note that the crystallite size for Ni$_{3.4}$Sn$_4$ is much smaller for Si$_C$-NiSn (7 ± 2 nm) than for Si$_O$-NiSn (39 ± 3 nm).

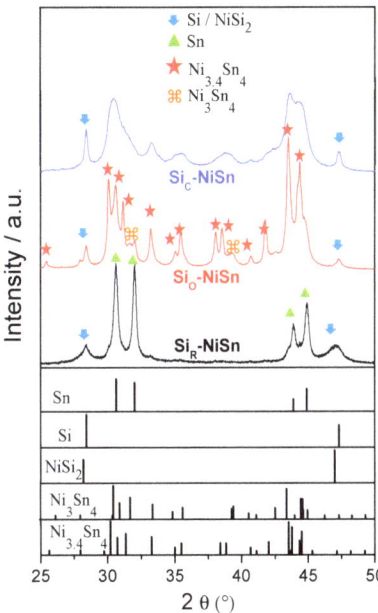

Figure 2. X-Ray Diffraction (XRD) patterns of the composites Si$_R$-NiSn, Si$_C$-NiSn and Si$_O$-NiSn after 20 h of milling. Position of diffraction lines for Sn, Si, NiSi$_2$, Ni$_3$Sn$_4$ and Ni$_{3.4}$Sn$_4$ phases as reported in Pearson's crystal data base [40] are shown in the bottom part of the figure.

Table 2. Crystallographic data for the Si_R-NiSn, Si_C-NiSn and Si_O-NiSn composites after 20 h of milling as determined from Rietveld analysis. Ni over-stoichiometry (x) in $Ni_{3+x}Sn_4$ and crystallite size (L in nm) for all phases are given. Standard deviations refereed to the last digit are given in parenthesis.

Sample	Phases	Content (wt.%)	S.G.	Cell Parameters				x in $Ni_{3+x}Sn_4$	L (nm)	R_B	R_{wp}
				a (Å)	b (Å)	c (Å)	β(°)				
Si_R-NiSn	Ni_3Sn_4	9(1)	C2/m	12.199 *	4.0609 *	5.2238 *	105.17 *	0*	10 *	6.8	9.7
	Si	13(1)	Fd-3m	5.430 *					15(2)	2.5	
	Sn	43(1)	$I4_1/amd$	5.8303(2)		3.1822(1)			27(1)	2.7	
	$NiSi_2$	35(2)	Fm-3m	5.4731 (5)					5(1)	3.6	
Si_C-NiSn	$Ni_{3.4}Sn_4$	85(2)	C2/m	12.357 (3)	4.060(1)	5.201(2)	104.31(2)	0.45(9)	7(2)	2.4	4.8
	Si	12(1)	Fd-3m	5.431(1)					30(2)	5.2	
	Sn	2(1)	$I4_1/amd$	5.8303 *		3.1822 *			27 *	4.5	
	$NiSi_2$	1(1)	Fm-3m	5.4731 *					5 *	2.4	
Si_O-NiSn	$Ni_{3.4}Sn_4$	49(2)	C2/m	12.448(2)	4.079(1)	5.209(1)	103.62(1)	0.4 *	39(3)	7.1	7.8
	Ni_3Sn_4	34(2)	C2/m	12.248(3)	4.046(1)	5.201(1)	104.88(1)	0 *	14(2)	5.9	
	Si	12(2)	Fd-3m	5.433(2)					19(2)	13.3	
	Sn	4(1)	$I4_1/amd$	5.8303 *		3.1822 *			27 *	4.7	
	$NiSi_2$	1(1)	Fm-3m	5.4731 *					5 *	17.3	

* fixed values.

The morphology of the three composites after 20 h of milling was examined using SEM and is displayed in Figure 3. The composite Si_R-NiSn consists of micrometric-size round-shaped particles (Figure 3a). The composite particles contain phase domains of dark tonality attributed to silicon nanoparticles [30] embedded in a light-grey matrix which is chemically homogeneous at the spatial resolution (~50 nm) of the BSE analysis (Figure 3b). In the case of material ground with Si_C, SEM-SE analysis (Figure 3c) shows that the composite particles are in the form of micrometer-sized platelets. SEM-BSE analysis (Figure 3d) reveals that the platelets are formed by particles with dark tonality (attributed to silicon) surrounded, as for the previous composite, by a chemically homogeneous light-grey matrix. Note that the silicon particle size (dark domains) is comparable for Si_R-NiSn and Si_C-NiSn composites. There are also brighter areas attributed to some $Ni_{3+x}Sn_4$ domains of micrometric size. Figure 3e,f show the SEM images for Si_O-NiSn composite. The composite particles also form platelets in the micrometric range. In the BSE-SEM image (Figure 3f), it is observed that the phase distribution within the particles is very inhomogeneous. There are very dark areas attributed to agglomerates of silicon particles and other areas with two different grayscales ascribed to the intermetallic $Ni_{3+x}Sn_4$ phase.

Figure 3. Scanning Electron Microscopy (SEM) images for Si_R-NiSn (**a,b**) Si_C-NiSn (**c,d**) and Si_O-NiSn (**e,f**) composites. Images were taken in either Secondary Electrons (SE) (**a,c,e**) or Back-Scattered Electrons (BSE) (**b,d,f**) modes.

To get a more accurate analysis of the chemically-homogeneous matrix in Si_R-NiSn and Si_C-NiSn composites, TEM analyses were performed (Figure 4). For the composite Si_R-NiSn (Figure 4, top), Si nanoparticles are surrounded by all elements. There is not a complete spatial correlation between Ni and Sn signals, which corroborates the decomposition of $Ni_{3.4}Sn_4$ as observed using XRD (Figure 2), leading to the formation of free Sn at the nanoscale. The analysis of the Si_C-NiSn composite (Figure 4, bottom) shows that the silicon particles are surrounded by a homogeneous matrix that contains Ni, Sn, C and Al. The Ni and Sn signals are spatially correlated indicating the presence of the Ni-Sn intermetallic at the nanometer scale in agreement with XRD results (Figure 2, Table 2). No preferential distribution of carbon is seen around the silicon particles: the carbon layer may have been dissolved upon grinding. However, carbon mapping should be considered with caution as carbon deposition is likely to occur under the electron beam. Complementary high-resolution TEM analysis (Figure S5) confirms the size of the coherent domains calculated by Rietveld refinement for the two main phases: about 30 nm for silicon (red area) and 8 nm for the Ni-Sn phase (black area).

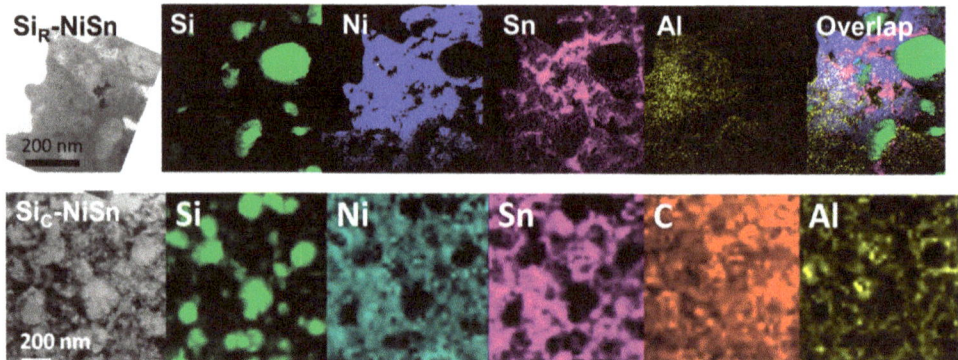

Figure 4. TEM images and elemental mapping for Si$_R$-NiSn ((**top**), Electron Energy Filtered Transmission Electron Microscopy (EFTEM) mapping) Si$_C$-NiSn ((**bottom**), STEM-EDX mapping).

3.2. Electrochemical Characterization

Figure 5 displays the potential profiles of the three studied Si-NiSn composites for the first and third cycles. At the first cycle, discharge (lithiation) profiles show a shoulder at 1.25 V attributed to the formation of the Solid Electrolyte Interface (SEI). Then, the potential profiles gradually decrease down to 0 V, showing several steps (~0.65, 0.40 and 0.35 V) for Si$_R$-NiSn, while the coated Si$_C$-NiSn and Si$_O$-NiSn composites have smooth potential profiles. Among the three composites, Si$_C$-NiSn has the lowest polarization potential. The first lithiation capacity is much lower for the oxide-coated Si$_O$-NiSn (685 mAh/g) than for Si$_R$-NiSn and Si$_C$-NiSn (950 and 1195 mAh/g, respectively). On charge (delithiation), either smooth or staircase potential profiles are again observed for coated (Si$_C$-NiSn and Si$_O$-NiSn) and bare (Si$_R$-NiSn) composites, respectively. At the third cycle, potential profiles show no evidence of SEI formation, and similarly to the 1st cycle, they are smooth for the coated composites while the bare composite has a staircase potential profile.

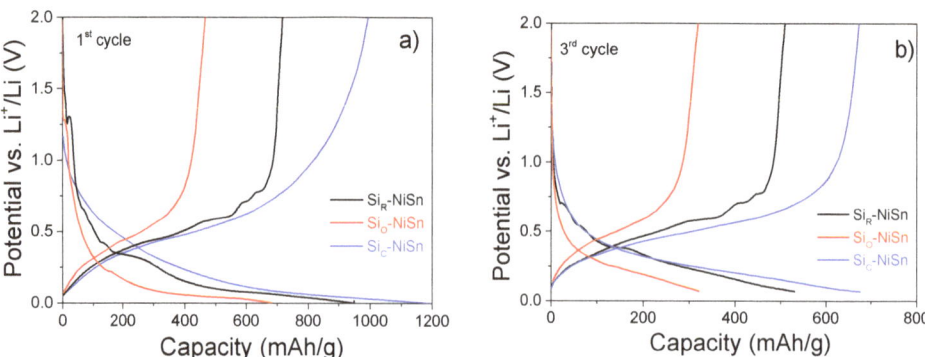

Figure 5. Discharge/charge profiles of Si$_R$-NiSn, Si$_O$-NiSn and Si$_C$-NiSn composites at the first (**a**) and third (**b**) galvanostatic cycle.

The evolution of reversible capacities (delithiation) and coulombic efficiency on cycling for the three composites is shown in Figure 6. For all composites, a significant capacity decay is observed during the first three activation cycles. Then, for the bare Si_R-NiSn composite, the capacity gradually decreases from 400 mAh/g at cycle 25 down to 210 mAh/g at cycle 200. In contrast, for the coated composites the capacity remains stable on cycling after activation, being significantly higher for Si_C-NiSn than for Si_O-NiSn. After 400 cycles, their reversible capacities are 505 and 215 mAh/g, respectively. As for the coulombic efficiency, ε_c (Figure 6b), it strongly depends on the composite at the first cycle. It ranges between 68% for oxide-coated Si_O-NiSn and 83% for the carbon-coated Si_C-NiSn composite. For the next cycles, the coulombic efficiencies drastically increase for all composites with typical values above 99.5%.

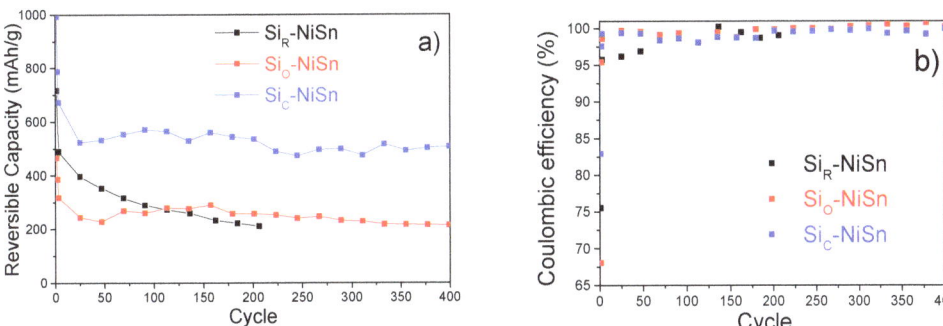

Figure 6. Evolution of the specific reversible capacity (**a**) and coulombic efficiency (**b**) of Si_R-NiSn, Si_O-NiSn and Si_C-NiSn composites during galvanostatic cycling.

To summarize, from the three studied composites, Si_C-NiSn exhibits the best electrochemical properties with a reversible capacity exceeding 500 mAh/g over 400 cycles. It has a reasonable initial coulombic efficiency of 83%, which increases to an average value of 99.6% between reference cycles 3 to 400.

4. Discussion

In this work, silicon nanoparticles with bare and chemically modified surfaces (with either C or O) have been used to prepare composites of Si-$Ni_{3.4}Sn_4$-Al using mechanical milling. Bare Si nanoparticles used as reference, Si_R, have an average size of 180 nm and contain minor traces of oxygen at the surface. The surface-modified Si-particles are nanometric (around 50–70 nm) and exhibit a core-shell structure with a shell thickness of 10 and 6 nm for Si_C and Si_O, respectively. Chemical analyses show that, for Si_C, the shell is mainly composed of carbon representing 30 wt.%C, whereas the shell chemistry of Si_O is identified as a silicon oxide with 46 wt.% of SiO_2, i.e., 25 wt.% of oxygen.

The chemical, microstructural and electrochemical properties of the Si-NiSn composites prepared with the different types of Si nanoparticles are summarized in Table 3. The ground composites consist of micrometer-sized particles that are round for bare silicon, Si_R, but turn to be platelet-like when using coated silicon Si_O and Si_C. Differences in composite morphology are ascribed to the fact that the Si surface chemistry plays a major role in the phase composition upon milling. Indeed, for coated Si, the major phase is the intermetallic $Ni_{3+x}Sn_4$, while Sn is the major phase when using bare Si. Formation of tin on milling results from the reaction

$$6Si + Ni_3Sn_4 \xrightarrow{milling} 3NiSi_2 + 4Sn \qquad (R1)$$

which produces $NiSi_2$ in addition as a secondary phase. The stoichiometry of the $Ni_{3+x}Sn_4$ phase is not considered here for the sake of simplicity. The occurrence of ductile Sn favors the formation of round-shaped particles on milling [41].

Table 3. Summary of the chemical, microstructural and electrochemical properties of the three composites Si$_R$-NiSn, Si$_C$-NiSn and Si$_O$-NiSn.

Composites	Si$_R$-NiSn	Si$_C$-NiSn	Si$_O$-NiSn
Composition (wt.%)	Si$_{0.22}$Ni$_{0.22}$Sn$_{0.53}$Al$_{0.03}$	Si$_{0.15}$Ni$_{0.22}$Sn$_{0.53}$Al$_{0.03}$C$_{0.07}$	Si$_{0.17}$Ni$_{0.22}$Sn$_{0.53}$Al$_{0.03}$O$_{0.06}$
Composition (at.%)	Si$_{0.46}$Ni$_{0.22}$Sn$_{0.26}$Al$_{0.06}$	Si$_{0.26}$Ni$_{0.18}$Sn$_{0.22}$Al$_{0.06}$C$_{0.28}$	Si$_{0.32}$Ni$_{0.20}$Sn$_{0.24}$Al$_{0.05}$O$_{0.19}$
Main phase (XRD; wt.% ±x)	Sn (43 ± 1)	Ni$_{3+x}$Sn$_4$ (85 ± 2)	Ni$_{3+x}$Sn$_4$ (83 ± 3)
Matrix phase distribution	Homogeneous	Homogeneous	Heterogenous
Particles morphology	Round-shaped	Platelets	Platelets
Sn phase (XRD; wt.% ±x)	43 ± 1	2 ± 1	4 ± 1
Crystal size Ni$_{3+x}$Sn$_4$ (nm)	9	7	14–39
Crystal size Si (nm)	15 ± 2	30 ± 2	19 ± 2
Potential profiles	Staircase	Smooth	Smooth
C$_{rev}$ (1st cyc.; mAhg^{-1})	715	995	465
C$_{rev}$ (3rd cyc.; mAhg^{-1})	490	675	320
C$_{rev}$ (mAhg^{-1})@cycle#	210@cycle200	505@cycle400	215@cycle400
ε_C (1st cycle; %)	75	83	68
ε_C (aver. 3–400 cycles; %)	-	99.6	99.8

Clearly, Si coating minimizes reaction R1, preserving the initial reactants, Si and Ni$_{3.4}$Sn$_4$, by avoiding direct contact between the two phases on milling. However, reaction R1 is not fully suppressed on prolonged milling since 2 and 4 wt.% of Sn are detected using XRD for Si$_C$ and Si$_O$, respectively. This reveals that carbon coating is more efficient than the oxide one, which is further supported by the fact that the stoichiometry of Ni$_{3+x}$Sn$_4$ remains constant for the Si$_C$-NiSn composite while it is partially altered for Si$_O$-NiSn (Table 2) [38]. The lower efficiency of the oxide coating to minimize the reaction between Si and Ni$_{3.4}$Sn$_4$ can be tentatively attributed to its small thickness (6 nm) and to the fact that the coating also contains Si in the form of SiO$_2$. Interestingly, it should also be noted that carbon coating not only minimizes Sn formation but also enhances the nanostructuration of the Ni$_{3+x}$Sn$_4$ phase. The crystallite size of Ni$_{3+x}$Sn$_4$ is much smaller with the carbon coating (crystallite size $L \sim 7$ nm) than for the oxide one ($L \sim 14$–39 nm). Thus, carbon coating allows efficient nanostructuration of the matrix leading to good chemical homogeneity around the Si nanoparticles (Figure 3).

The difference in chemical and microstructural properties between Si$_C$-NiSn, Si$_O$-NiSn and Si$_R$-NiSn composites lead to clearly distinct electrochemical behaviors, which are also summarized in Table 3. The reference Si$_R$-NiSn composite has staircase potential profiles, moderate initial capacity and poor cycle-life. Oxide coating of Si nanoparticles leads to smooth potential profiles and good cycle-life but at the expense of limited capacity. Finally, carbon coating not only lead to smooth potential profiles but also to high capacity and coulombic efficiency over hundreds of cycles. Smooth profiles are preferred to staircase ones since the volume changes of the active materials induced by phase transformations occur gradually for the former, minimizing mechanical degradation on cycling. A better insight into the different electrochemical properties between the composites can be gained at the light of the microstructural properties and through deep analysis of potential profiles (Figure 5) by evaluating Differential Capacity Plots (DCPs).

Figure 7 displays the DCP plots for the three composites at the 1st, 3rd and 400 cycles. For the first galvanostatic cycle of the bare Si$_R$-NiSn composite (Figure 7a), four clear reduction peaks of moderate intensity are observed at 0.66, 0.56, 0.42 and 0.34 V and a broad additional one below 0.1 V. The first four peaks are attributed to lithiation of the main phase (free Sn), whereas the latter one is assigned to the formation of Li-rich Li$_y$Si and Li$_z$Sn alloys [31]. In the anodic branch, four clear oxidation peaks are observed at 0.44, 0.58, 0.70 and 0.78 V, which are attributed to the decomposition of the different Li$_z$Sn alloys (Li$_7$Sn$_2$, Li$_5$Sn$_2$, LiSn and Li$_2$Sn$_5$) in agreement with previous reports [38,42]. The signal at 0.58 V is in fact a triplet due to the decomposition of three Li$_z$Sn alloys of close composition: Li$_{13}$Sn$_5$, Li$_5$Sn$_2$ and Li$_7$Sn$_3$ [42,43]. In addition, an anodic bump and a broad oxidation peak can be observed at 0.32 and 0.46 V, which are attributed to the decomposition of amorphous

Li$_{3.16}$Si and Li$_7$Si$_3$ alloys, respectively [20,38,44]. The detected dQ/dV peaks for the first galvanostatic cycle in the bare Si$_R$-NiSn composite are consistent with the coexistence of pure Si and Sn phases (Table 2). The width of the peaks is anticorrelated with the crystallinity of the phases: anodic peaks due to decomposition of amorphous Li$_y$Si alloys formed during the first composite lithiation [45] are wider than those of the crystalline Li$_z$Sn ones [46]. Interestingly, at the third cycle (Figure 7b), dQ/dV peaks attributed to the formation and decomposition of Li$_z$Sn alloys are sharper than those of the first cycle, which suggests the coarsening of Sn domains on cycling [47]. This agglomeration favorizes discrete volume changes, leading to electrode cracking [19] and severe capacity decay for the bare Si$_R$-NiSn composite, as observed in Figure 6a.

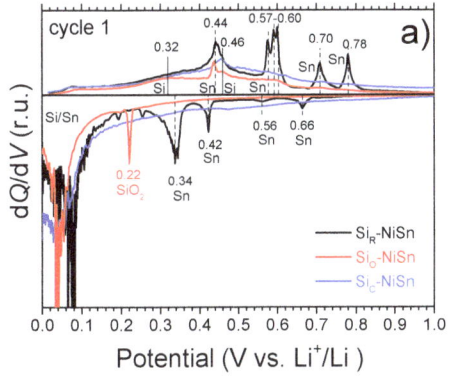
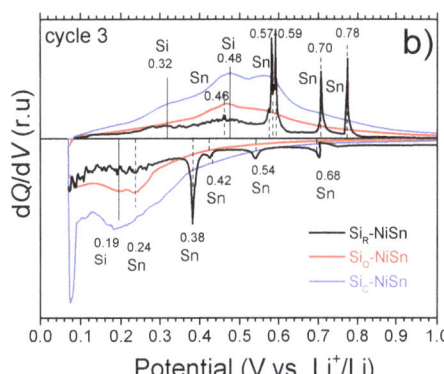
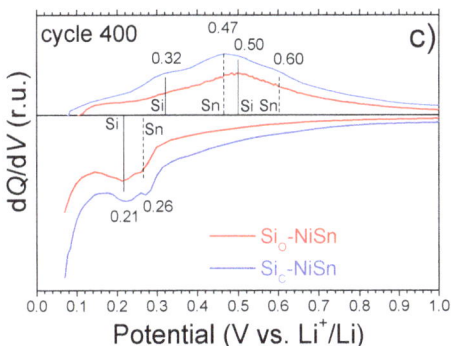

Figure 7. Differential capacity plots for all composites at cycles (**a**) 1, (**b**) 3 and (**c**) 400. Solid and dashed vertical lines have been used to identify formation/decomposition of Li$_y$Si and Li$_z$Sn alloys, respectively.

The dQ/dV plot for the oxide-coated Si$_O$-NiSn composite at the 1st cycle (Figure 7a) displays two reduction peaks at 0.22 and 0.04 V which are assigned to the lithiation of silicon oxide SiO$_2$ and crystalline Si, respectively [45,48]. The anodic branch at the 1st cycle is almost featureless with a broad peak at 0.46 V and a shoulder at ~0.59 V attributed to decomposition of Li$_z$Sn alloys, as well as two shoulders at 0.32 and 0.46 V assigned to decomposition of the Li$_y$Si ones. The broadness of these signals is a signature of the low crystallinity of the reacting phases for this composite. This strongly differs from the well-defined oxidation peaks observed for Si$_R$-NiSn, which evidences a different chemical and microstructural state of Sn between the Si$_R$-NiSn and Si$_O$-NiSn composites. For Si$_R$-NiSn, free Sn is formed during ball milling (Table 2) with a crystallite size of 27 nm. As mentioned before, free Sn likely coarsens due to agglomeration during electrochemical cycling. In contrast, for the Si_O-NiSn composite, Sn remains alloyed with Ni in the form of nanometric Ni_3Sn_4 intermetallic after ball milling. Upon lithiation/delithiation, Ni_3Sn_4 undergoes a reversible conversion reaction which can be expressed as [23,38,49]

$$Ni_3Sn_4 + 14Li \leftrightarrow 2Li_7Si_2 + 3Ni \quad \text{(R2)}$$

This conversion reaction ensures the nanostructured state of Sn-forming alloys, Ni_3Sn_4 and Li_zSn, on cycling [12]. It should be also noticed that very similar featureless anodic branches are observed at cycles 3 (Figure 7b) and 400 (Figure 7c), which is concomitant with the long-term cycling stability of Si_O-NiSn (Figure 6a). As for the cathodic branch at cycles 3 and 400, two broad peaks are detected at 0.24 and 0.19 V that are tentatively attributed to the formation of poorly crystallized Li_zSn and Li_ySi alloys, respectively [31]. The sharp peak detected at 0.22 V at the 1st reduction attributed to the SiO_2 lithiation is not detected in subsequent cycles showing its irreversible behavior. Indeed, as reported by Guo et al. [48], SiO_2 can react with lithium through two reaction paths:

$$SiO_2 + 4Li^+ + 4e^- \rightarrow 2Li_2O + Si \quad \text{(R3)}$$

$$2SiO_2 + 4Li^+ + 4e^- \rightarrow Li_4SiO_4 + Si \quad \text{(R4)}$$

The irreversibility of these reactions accounts for the low coulombic efficiency of Si_O-NiSn in the first cycle (68%, Figure 6b) [50] and explains the low reversible capacity of this composite. In addition, the effect of the poor chemical homogeneity of the composite matrix (Figure 3f) on the limited first lithiation capacity (685 mAh/g) of the Si_O-NiSn composite cannot be ruled out (Figure 5a).

Finally, the dQ/dV plots for the carbon-coated Si_C-NiSn composite exhibit, as a general trend, smooth traces both for cathodic and anodic branches and all over the 400 cycles (Figure 7). At the first reduction, a unique clear peak below 0.1 V is detected and attributed to formation of Li-rich Li_ySi alloys and the conversion reaction R2 for the major Ni_3Sn_4 phase. No signal of large SEI formation is observed, which concurs with the high initial coulombic efficiency of this composite (87%). At the 1st oxidation, a broad peak at 0.46 V is attributed to decomposition of Li_ySi alloys, while bumps at ~0.44 and 0.58 V can be assigned to decomposition of Li_zSn alloys leading to the recovery of Ni_3Sn_4 [31]. At cycles 3 and 400, very similar and smooth curves are detected showing several bumps that point out good and stable reversibility in the lithiation of Si and Ni_3Sn_4 counterparts of the Si_C-NiSn composite. It should be noted that the area under the dQ/dV plots is much larger for Si_C-NiSn than for Si_O-NiSn showing the higher capacity of the former (Figure 6).

5. Conclusions

Modification of the Si surface chemistry clearly affects the chemical and microstructural properties of Si-$Ni_{3.4}Sn_4$-Al composites as anode materials in Li-ion cells. First, it plays a protective role in the mechanochemical synthesis of the composite. This is indeed an effective solution for limiting the reaction between silicon and $Ni_{3.4}Sn_4$ during grinding and thus preventing the formation of detrimental free Sn. The milling process with Si-coated particles leads to a platelet-like morphology of the composites for both oxide- and carbon-coated silicon in contrast with the round-shaped one using bare silicon. However, differences in the microstructure of the composite matrix are found as a function of the surface chemistry, being chemically heterogeneous at the nanoscale for oxide coating while it is homogeneous for the carbon one. This leads to a low lithiation capacity for oxide coating and, moreover, low coulombic efficiency at the first cycle due to an irreversible reaction between SiO_2 and lithium. The use of carbon coating leads to a homogeneous matrix surrounding Si nanoparticles leading to a high reversible capacity that keeps stable after hundreds of cycles. Such an approach allows high performance materials usable as anodes for high energy density batteries to be developed.

Supplementary Materials: The following are available online at https://www.mdpi.com/2079-4991/11/1/18/s1, Figure S1: Microstructural characterization of bare Si powder, Figure S2: Microstructural characterization of carbon-coated Si powder, Figure S3: Microstructural characterization of oxide-coated Si powder, Figure S4: Evolution of the XRD patterns of Si-NiSn composites as a function of milling time, Figure S5: Graphical output of Rietveld analysis of Si-NiSn composites, Figure S6: High-resolution TEM image of the Si$_C$-NiSn composite.

Author Contributions: Conceptualization, T.A., C.T., F.C., C.J., N.M., M.L.; methodology, C.T., F.C., E.L., M.L.; validation, F.C., C.J., N.M., M.L.; investigation, T.A., C.T., F.C., M.L.; writing—original draft preparation, T.A., C.T., F.C., M.L.; writing—review and editing, T.A., C.T., F.C., E.L., C.J., N.M., M.L.; supervision, F.C., M.L.; project administration, F.C.; funding acquisition, F.C., M.L., C.J., N.M. All authors have read and agreed to the published version of the manuscript.

Funding: This research was funded by the the French Research Agency (ANR), project NEWMASTE, grant number n° ANR-13-PRGE-0010.

Acknowledgments: The authors are grateful to Remy Pires for SEM and EDX analysis and Valérie Lalanne for TEM sample preparation.

Conflicts of Interest: The authors declare no conflict of interest.

References

1. Winter, M.; Barnett, B.; Xu, K. Before Li Ion Batteries. *Chem. Rev.* **2018**, *118*, 11433–11456. [CrossRef] [PubMed]
2. El Kharbachi, A.; Zavorotynska, O.; Latroche, M.; Cuevas, F.; Yartys, V.; Fichtner, M. Exploits, advances and challenges benefiting beyond Li-ion battery technologies. *J. Alloys Compd.* **2020**, *817*, 153261. [CrossRef]
3. Goriparti, S.; Miele, E.; De Angelis, F.; Di Fabrizio, E.; Zaccaria, R.P.; Capiglia, C. Review on recent progress of nanostructured anode materials for Li-ion batteries. *J. Power Sources* **2014**, *257*, 421–443. [CrossRef]
4. Wang, A.; Kadam, S.; Li, H.; Shi, S.; Qi, Y. Review on modeling of the anode solid electrolyte interphase (SEI) for lithium-ion batteries. *NPJ Comput. Mater.* **2018**, *4*, 1–26. [CrossRef]
5. Yan, Y.; Li, C.; Liu, C.; Mutlu, Z.; Dong, B.; Liu, J.; Ozkan, C.S.; Ozkan, M. Bundled and dispersed carbon nanotube assemblies on graphite superstructures as free-standing lithium-ion battery anodes. *Carbon* **2019**, *142*, 238–244. [CrossRef]
6. Yu, K.; Wang, J.; Wang, X.; Liang, J.-C.; Liang, C. Sustainable application of biomass by-products: Corn straw-derived porous carbon nanospheres using as anode materials for lithium ion batteries. *Mater. Chem. Phys.* **2020**, *243*, 122644. [CrossRef]
7. Zhong, M.; Yan, J.; Wu, H.; Shen, W.; Zhang, J.; Yu, C.; Li, L.; Hao, Q.; Gao, F.; Tian, Y.; et al. Multilayer graphene spheres generated from anthracite and semi-coke as anode materials for lithium-ion batteries. *Fuel Process. Technol.* **2020**, *198*, 106241. [CrossRef]
8. Zeng, H.; Xinga, B.; Zhang, C.; Chen, L.; Zhao, H.; Han, X.; Yi, G.; Huang, G.; Zhang, C.; Cao, Y. In Situ Synthesis of MnO$_2$/Porous Graphitic Carbon Composites as High-Capacity Anode Materials for Lithium-Ion Batteries. *Energy Fuels* **2020**, *34*, 2480–2491. [CrossRef]
9. Lu, Y.; Yu, L.; Lou, X.W. Nanostructured Conversion-type Anode Materials for Advanced Lithium-Ion Batteries. *Chem* **2018**, *4*, 972–996. [CrossRef]
10. Sartori, S.; Cuevas, F.; Latroche, M. Metal hydrides used as negative electrode materials for Li-ion batteries. *Appl. Phys. A* **2016**, *122*, 135. [CrossRef]
11. Feng, K.; Li, M.; Liu, W.; Kashkooli, A.G.; Xiao, X.; Cai, M.; Chen, Z. Silicon-Based Anodes for Lithium-Ion Batteries: From Fundamentals to Practical Applications. *Small* **2018**, *14*, 1702737. [CrossRef] [PubMed]
12. Xin, F.; Whittingham, M.S. Challenges and Development of Tin-Based Anode with High Volumetric Capacity for Li-Ion Batteries. *Electrochem. Energy Rev.* **2020**, *3*, 643–655. [CrossRef]
13. Liang, B.; Liu, Y.; Xu, Y. Silicon-based materials as high capacity anodes for next generation lithium ion batteries. *J. Power Sources* **2014**, *267*, 469–490. [CrossRef]
14. Nguyen, T.L.; Kim, D.S.; Hur, J.; Park, M.S.; Kim, I.T. Ni-Sn-based hybrid composite anodes for high-performance lithium-ion batteries. *Electrochim. Acta* **2018**, *278*, 25–32. [CrossRef]
15. Larcher, D.; Beattie, S.; Morcrette, M.; Edström, K.; Jumasc, J.-C.; Tarascona, J.-M. Recent findings and prospects in the field of pure metals as negative electrodes for Li-ion batteries. *J. Mater. Chem.* **2007**, *17*, 3759–3772. [CrossRef]
16. Binder, L.O. Metallic Negatives. In *Handbook of Battery Materials*; Wiley: Weinheim, Germany, 1998; pp. 195–208.
17. Obrovac, M.N.; Christensen, L. Structural Changes in Silicon Anodes during Lithium Insertion/Extraction. *Electrochem. Solid-State Lett.* **2004**, *7*, A93–A96. [CrossRef]
18. Kasavajjula, U.; Wang, C.; Appleby, A.J. Nano- and bulk-silicon-based insertion anodes for lithium-ion secondary cells. *J. Power Sources* **2007**, *163*, 1003–1039. [CrossRef]
19. Beaulieu, L.Y.; Eberman, K.W.; Turner, R.L.; Krause, L.J.; Dahn, J.R. Colossal Reversible Volume Changes in Lithium Alloys. *Electrochem. Solid-State Lett.* **2001**, *4*, A137–A140. [CrossRef]
20. Jerliu, B.; Hüger, E.; Dörrer, L.; Seidlhofer, B.K.; Steitz, R.; Horisberge, M.; Schmidt, H. Lithium insertion into silicon electrodes studied by cyclic voltammetry and operando neutron reflectometry. *Phys. Chem. Chem. Phys.* **2018**, *20*, 23480–23491. [CrossRef]
21. Szczech, J.R.; Jin, S. Nanostructured silicon for high capacity lithium battery anodes. *Energy Environ. Sci.* **2011**, *4*, 56–72. [CrossRef]

22. Song, T.; Kil, K.C.; Jeon, Y.; Lee, S.; Shin, W.C.; Chung, B.; Kwon, K.; Paik, U. Nitridated Si–Ti–Ni alloy as an anode for Li rechargeable batteries. *J. Power Sources* **2014**, *253*, 282–286. [CrossRef]
23. Edfouf, Z.; Fariaut-Georges, C.; Cuevas, F.; Latroche, M.; Hézèque, T.; Caillon, G.; Jordy, C.; Sougrati, M.; Jumas, J. Nanostructured $Ni_{3.5}Sn_4$ intermetallic compound: An efficient buffering material for Si-containing composite anodes in lithium ion batteries. *Electrochim. Acta* **2013**, *89*, 365–371. [CrossRef]
24. Mukaibo, H.; Momma, T.; Osaka, T. Changes of electro-deposited Sn–Ni alloy thin film for lithium ion battery anodes during charge discharge cycling. *J. Power Sources* **2005**, *146*, 457–463. [CrossRef]
25. Naille, S.; Dedryvère, R.; Zitoun, D.; Lippens, P. Atomic-scale characterization of tin-based intermetallic anodes. *J. Power Sources* **2009**, *189*, 806–808. [CrossRef]
26. Kitada, A.; Fukuda, N.; Ichii, T.; Sugimura, H.; Murase, K. Lithiation behavior of single-phase Cu–Sn intermetallics and effects on their negative-electrode properties. *Electrochim. Acta* **2013**, *98*, 239–243. [CrossRef]
27. Alcántara, R.; Ortiz, G.; Rodríguez, I.; Tirado, J.L. Effects of heteroatoms and nanosize on tin-based electrodes. *J. Power Sources* **2009**, *189*, 309–314. [CrossRef]
28. Wang, Z.; Tian, W.; Liu, X.; Li, Y.; Li, X. Nanosized Si–Ni alloys anode prepared by hydrogen plasma–metal reaction for secondary lithium batteries. *Mater. Chem. Phys.* **2006**, *100*, 92–97. [CrossRef]
29. Lee, J.-S.; Shin, M.-S.; Lee, S.-M. Electrochemical properties of polydopamine coated Ti-Si alloy anodes for Li-ion batteries. *Electrochim. Acta* **2016**, *222*, 1200–1209. [CrossRef]
30. Azib, T.; Thaury, C.; Fariaut-Georges, C.; Hézèque, T.; Cuevas, F.; Jordy, C.; Latroche, M. Role of silicon and carbon on the structural and electrochemical properties of $Si-Ni_{3.4}Sn_4$-Al-C anodes for Li-ion batteries. *Mater. Today Commun.* **2020**, *23*, 101160. [CrossRef]
31. Edfouf, Z.; Sougrati, M.; Fariaut-Georges, C.; Cuevas, F.; Jumas, J.-C.; Hézèque, T.; Jordy, C.; Caillon, G.; Latroche, M. Reactivity assessment of lithium with the different components of novel $Si/Ni_{3.4}Sn_4$/Al/C composite anode for Li-ion batteries. *J. Power Sources* **2013**, *238*, 210–217. [CrossRef]
32. Edfouf, Z.; Cuevas, F.; Latroche, M.; Georges, C.; Jordy, C.; Hézèque, T.; Caillon, G.; Jumas, J.; Sougrati, M. Nanostructured Si/Sn–Ni/C composite as negative electrode for Li-ion batteries. *J. Power Sources* **2011**, *196*, 4762–4768. [CrossRef]
33. Hatchard, T.D.; Topple, J.M.; Fleischauer, M.D.; Dahn, J.R. Electrochemical Performance of SiAlSn Films Prepared by Combinatorial Sputtering. *Electrochem. Solid-State Lett.* **2003**, *6*, A129–A132. [CrossRef]
34. Wang, M.-S.; Fan, L.-Z. Silicon/carbon nanocomposite pyrolyzed from phenolic resin as anode materials for lithium-ion batteries. *J. Power Sources* **2013**, *244*, 570–574. [CrossRef]
35. Yu, J.; Yang, J.; Feng, X.; Jia, H.; Wang, J.; Lu, W. Uniform Carbon Coating on Silicon Nanoparticles by Dynamic CVD Process for Electrochemical Lithium Storage. *Ind. Eng. Chem. Res.* **2014**, *53*, 12697–12704. [CrossRef]
36. Rodriguez-Carvajal, J. Fullprof: A Program for Rietveld Refinement and Pattern Matching Analysis. *Phys. B* **1993**, *192*, 55–69.
37. Xu, K. Nonaqueous Liquid Electrolytes for Lithium-Based Rechargeable Batteries. *Chem. Rev.* **2004**, *104*, 4303–4418. [CrossRef]
38. Azib, T.; Bibent, N.; Latroche, M.; Fischer, F.; Jumas, J.-C.; Olivier-Fourcade, J.; Jordy, C.; Lippens, P.-E.; Cuevas, F. Ni–Sn intermetallics as an efficient buffering matrix of Si anodes in Li-ion batteries. *J. Mater. Chem. A* **2020**. [CrossRef]
39. Hubbard, C.R.; Swanson, H.E.; Mauer, F.A. A silicon powder diffraction standard reference material. *J. Appl. Crystallogr.* **1975**, *8*, 45–48. [CrossRef]
40. Villar, P.; Cenzual, K. *Pearson's Crystal Data: Crystal Structure Database for Inorganics Compounds, Release 2010/2011*; ASM International: Materials Park, OH, USA, 2010; Available online: http://www.crystalimpact.com/pcd/ (accessed on 3 November 2020).
41. Huang, J.; Wu, Y.; Ye, H. Ball milling of ductile metals. *Mater. Sci. Eng. A* **1995**, *199*, 165–172. [CrossRef]
42. Wang, C.; Appleby, A.J.; Little, F.E. Electrochemical study on nano-Sn, $Li_{4.4}Sn$ and $AlSi_{0.1}$ powders used as secondary lithium battery anodes. *J. Power Sources* **2001**, *93*, 174–185. [CrossRef]
43. Wen, C.J.; Huggins, R.A. Thermodynamic Study of the Lithium-Tin System. *J. Electrochem. Soc.* **1981**, *128*, 1181–1187. [CrossRef]
44. Jiménez, A.R.; Klöpsch, R.; Wagner, R.; Rodehorst, U.C.; Kolek, M.; Nölle, R.; Winter, M.; Placke, T. A Step toward High-Energy Silicon-Based Thin Film Lithium Ion Batteries. *ACS Nano* **2017**, *11*, 4731–4744. [CrossRef] [PubMed]
45. Li, J.; Dahn, J.R. An In Situ X-Ray Diffraction Study of the Reaction of Li with Crystalline Si. *J. Electrochem. Soc.* **2007**, *154*, A156–A161. [CrossRef]
46. Liang, J.; Lu, Y.; Liu, J.; Liu, X.; Gong, M.; Deng, S.; Yang, H.; Liu, P.; Wanga, D. Oxides overlayer confined Ni_3Sn_2 alloy enable enhanced lithium storage performance. *J. Power Sources* **2019**, *441*, 227185. [CrossRef]
47. Winter, M.; Besenhard, J.O. Electrochemical lithiation of tin and tin-based intermetallics and composites. *Electrochim. Acta* **1999**, *45*, 31–50. [CrossRef]
48. Guo, B.; Shu, J.; Wang, Z.; Yang, H.; Shi, L.; Liu, Y.; Chen, L. Electrochemical reduction of nano-SiO_2 in hard carbon as anode material for lithium ion batteries. *Electrochem. Commun.* **2008**, *10*, 1876–1878. [CrossRef]
49. Naille, S.; Lippens, P.E.; Morato, F.; Olivier-Fourcade, J. 119Sn Mössbauer study of nickel–tin anodes for rechargeable lithium-ion batteries. *Hyperfine Interact.* **2006**, *167*, 785–790. [CrossRef]
50. Tu, J.; Yuan, Y.; Zhan, P.; Jiao, H.; Wang, X.; Zhu, H.; Jiao, S. Straightforward Approach toward SiO_2 Nanospheres and Their Superior Lithium Storage Performance. *J. Phys. Chem. C* **2014**, *118*, 7357–7362. [CrossRef]

Article

Highly Porous Free-Standing rGO/SnO$_2$ Pseudocapacitive Cathodes for High-Rate and Long-Cycling Al-Ion Batteries

Timotheus Jahnke [1], Leila Raafat [2], Daniel Hotz [1], Andrea Knöller [2,3], Achim Max Diem [2], Joachim Bill [2] and Zaklina Burghard [2,*]

[1] Max Planck Institute for Medical Research, 61920 Heidelberg, Germany; Timotheus.jahnke@mr.mpg.de (T.J.); hotz@mr.mpg.de (D.H.)
[2] Institute for Materials Science, University of Stuttgart, 70569 Stuttgart, Germany; leila.raafat@imw.uni-stuttgart.de (L.R.); andrea.knoeller@hahn.schickard.de (A.K.); diem@imw.uni-stuttgart.de (A.M.D.); bill@imw.uni-stuttgart.de (J.B.)
[3] Institute for Micro Assembly Technology of the Hahn-Schickard, 70569 Stuttgart, Germany
[*] Correspondence: zaklina.burghard@imw.uni-stuttgart.de; Tel.: +49-711-685-61958

Received: 14 September 2020; Accepted: 12 October 2020; Published: 14 October 2020

Abstract: Establishing energy storage systems beyond conventional lithium ion batteries requires the development of novel types of electrode materials. Such materials should be capable of accommodating ion species other than Li$^+$, and ideally, these ion species should be of multivalent nature, such as Al^{3+}. Along this line, we introduce a highly porous aerogel cathode composed of reduced graphene oxide, which is loaded with nanostructured SnO$_2$. This binder-free hybrid not only exhibits an outstanding mechanical performance, but also unites the pseudocapacity of the reduced graphene oxide and the electrochemical storage capacity of the SnO$_2$ nanoplatelets. Moreover, the combination of both materials gives rise to additional intercalation sites at their interface, further contributing to the total capacity of up to 16 mAh cm^{-3} at a charging rate of 2 C. The high porosity (99.9%) of the hybrid and the synergy of its components yield a cathode material for high-rate (up to 20 C) aluminum ion batteries, which exhibit an excellent cycling stability over 10,000 tested cycles. The electrode design proposed here has a great potential to meet future energy and power density demands for advanced energy storage devices.

Keywords: aluminum ion batteries; reduced graphene oxide; tin dioxide; 3D electrode materials; mechanical properties

1. Introduction

Electrochemical energy storage systems based on lithium ions are nowadays a well-established concept, which enables the application of a broad spectrum of technologies, ranging from microelectronics over portable electronic devices to even electrical cars. Such lithium ion batteries (LIBs) offer excellent energy density and long-term stability. However, the limited lithium resources paired with its high cost and safety hazards make LIBs only a medium-term solution. In the long term, sustainable alternatives, e.g., other metal ion batteries [1], are more favorable. Researchers have been investigating monovalent ion batteries, like sodium and potassium-based electrochemical systems as alternatives [2,3], due to the high abundance and cost efficiency of these elements. However, larger monovalent ions usually have a sluggish diffusivity and lower intercalation voltages at similar energy densities (one electron per ion transfer). To improve the energy density of an active material, electrochemical systems based on multivalent ion exchange have been in discussion among researchers [4,5]. Among them, aluminum-based electrochemical systems have become one of the most promising candidates, owing to the aluminum's natural abundance, low cost and inherent safety [6–9]. Moreover, the three-valent aluminum could significantly boost the energy storage capacity compared to the single-valent lithium [10].

Establishing aluminum ion batteries (AIBs) requires the development of suitable electrolytes and tailored electrode materials. Imidazole-based ionic liquids have already been identified as suitable electrolytes, as they exhibit a low inner resistance, high solubility of the aluminum salt $AlCl_3$ and good electrochemical stability [11]. The search for cathode materials, however, is still ongoing, and is among the most discussed current topics in this field [7,12]. One fundamental requirement for such a material is its capability to intercalate Al^{3+} ions or chloroaluminate ions at a relatively high operating potential [7]. In this respect, a promising candidate is graphite, with its layered structure that facilitates the access of intercalating ions and provides sufficient space for them. Specifically, it was demonstrated that graphite-based materials immersed in an imidazole-based electrolyte deliver storage capacities of up to 60 mAh g^{-1} at a high current density of up to 2 A g^{-1} [13]. Other carbon-based materials, such as carbon nanotubes or graphene/reduced graphene oxide (rGO), have likewise been tested as electrode material for AIBs [7,8,14–17]. A storage capacity between 60 to 150 mAh g^{-1} could be achieved due to the reversible intercalation of chloroaluminate ions. Regarding rGO electrodes, electrochemical characterization further revealed an extremely high charging capability of up to 20 A g^{-1}, delivering capacities stable over thousands of cycles [16]. This enhanced electrochemical performance could be ascribed to the material's pseudocapacitive behavior [18], which arises from the adsorption of chloroaluminate ions onto its surface. The micro/mesoporosity and large specific surface area thereby promote the pseudocapacity, thus enabling a full charging process in minutes or even seconds.

A substantial advantage of pseudocapacitive materials, such as rGO, is that the main contribution to their energy storage capability is attributed to the ion adsorption rather than the bulk ion-diffusion, i.e., de-/intercalation. However, in comparison to LIBs, the intercalation or adsorption of negatively charged aluminum complexes in graphitic materials is still relatively low concerning their gravimetric capacity (<150 mAh g^{-1}) and the energy density (<0.3 Wh g^{-1}) [16]. In addition to graphitic materials, electrochemically active materials, such as sulphur [19,20], sulphides [21–25], and oxides [26–28], were investigated for their applicability in AIBs. Even though they showed great potential in terms of their electrochemical storage capacity, they were often lacking in terms of cycling stability or electrical conductivity, leading to capacity fading and short cycle life [12]. Among the oxides, tin oxide (SnO_2) stands out due to its high specific capacity (434 mAh g^{-1}) [28], good electrical conductivity (1–100 S cm^{-1}) [29,30], and wide availability, although it has only been sparsely investigated. Nevertheless, the combination of carbon-based materials and SnO_2 has been reported and investigated for LIBs [31,32]. The integration of SnO_2 nanostructures into a flexible carbon-based matrix has an essential impact of the electrodes performance and can be achieved for example by thermal post treatment [33].

To unite the main properties of pseudocapacitors with those of batteries, thus presenting fast charging rates and good cycling stability with a high storage capacity, the fabrication of hybrid electrodes presents great potential. Moreover, tailoring the hybrid electrode's microstructure to maximize the specific surface area and shorten the diffusion paths could further boost the synergy of ion adsorption and intercalation. In this work, we therefore united pseudocapacitive rGO nanosheets with electrochemically active SnO_2 nanoplatelets in highly porous, binder-free aerogel electrodes. Their applicability as cathodes in novel AIBs was tested with respect to their mechanical and electrochemical performance. This work shows that tailoring the porosity and surface area of carbon-based electrode materials in combination with an electrochemically active material, enhances the mechanical stability and the pseudocapacitive performance in AIB.

2. Materials and Methods

2.1. Fabrication of the Single Components

Graphene oxide was synthesized employing a modified Hummers method [34] Here, 0.5 g graphite flakes (NGS Naturgraphit GmbH) and 0.6 g KNO_3 (Merck KGaA, Darmstadt, Germany) were added to 23 mL of H_2SO_4 (Merck KGaA, Darmstadt, Germany, 98%) in a three-neck, round bottom flask, which is placed in an ice bath. Constant stirring for 10 min was applied to ensure a good dispersion of the graphite flakes in the H_2SO_4, whereas afterwards, 3 g $KMnO_4$ (Merck KGaA, Darmstadt, Germany) was added, and the temperature adjusted to 35 °C and held for 6 h, while continuing the stirring. Subsequently, 40 mL of dd-H_2O was added dropwise, and the temperature was increased to 80 °C and held for 30 min. The reaction was interrupted by slowly adding 100 mL dd-H_2O and 3 mL H_2O_2 to the dispersion.

To remove the residual reagents and increase the highly acidic pH (<1), the GO dispersion was centrifuged (SORVALL RC6, Thermo Fischer Scientific, Schwerte, Germany) for 10 min at 17,000 relative centrifugal force. The transparent portion in the flask was decanted after centrifugation, dd-H_2O added, shortly stirred up and then centrifuged again. This process was repeated at least 10 times, until the pH was above four. The GO sheets, with a sheet size less than 1μm were obtained from IoLiTec, Ionic Liquid Technologies, Karlsruhe, Germany with an initial concentration of 5 mg mL^{-1}.

The SnO_2 nanoparticles were synthesized using a common hydrothermal approach with $SnCl_2$ salt and ammonia as initial reagents [35] 1.36 g $SnCl_2 * 2H_2O$ were dissolved in 10 mL dd-H_2O and 10 mL Ethanol (Merck KGaA, Darmstadt, Germany, p.a.) for 10 min. Subsequently, 24 mL of 0.55 M ammonia solution was added slowly mixture, which turned opaque yellow upon the addition of ammonia. Consequently, the mixture was transferred to a 25 mL Teflon-lined autoclave and hydrothermally treated for 6 h at 120 °C. The material was finally washed alternatively three times with ethanol and three times with dd-H_2O to obtain a clean powder.

2.2. Ice-Templating and Annealing

To obtain the hybrid composed of rGO and SnO_2, an aqueous dispersion of the respective GO sheets (66.6 wt%) was mixed with SnO_2 nanoparticles (33.3 wt%) and sonicated for 10 min in an ultrasonic bath with a 90 W power supply. The dispersion was then transferred into a PTFE mold, which was placed beforehand on the cold finger, and frozen by applying a unidirectional temperature gradient of 18.5 K mm^{-1} in z-direction. To this end, self-supporting aerogels, with a radius of 4 mm and a height of 8 mm and a porosity of >99.9% were obtained. These self-supporting aerogels were transferred into the freeze drier (L10E, Dieter Piatkowski, Petershausen, Germany), which was cooled down to −50 °C before evacuation. After the removal of the ice crystals, the aerogels were reduced in a quartz tube furnace in Argon (99.95 Ar), ramping with 4 K min^{-1} to 500 °C, at which they were kept for two hours. The obtained hybrid aerogels had a 41 wt% rGO to 59 wt% SnO_2 weight ratio and a weight around 1.55 mg ± 0.03 mg, whereas the pure rGO aerogels had a weight of 0.748 mg ± 0.002 mg, obtained from weight measurements on three samples each.

2.3. Microstructural Characterization

For microstructural investigations, scanning electron microscopy (Zeiss Ultra 5, Carl Zeiss AG, Oberkochen, Germany) was used. X-ray diffraction was performed on a Bruker D8 system (Bruker Cooperation, MA, USA) using copper K_α radiation in the range of 10–90° with 0.01° as step size. The crystallite size was calculated using the Scherrer equation with a form factor of 1 and a Full-Width-Half-Maximum of 1.87 theta around a Bragg Angle of 26.59 deg. Brunauer-Emmett-Teller (BET) tests were performed on a physisorption analyzer (Quantachrome Instruments Autosorb iQ3, Quantachrome/Anton Paar GmbH, Graz, Austria).

2.4. Mechanical Characterization

The mechanical measurements were conducted on a Bose Electroforce (TA Instruments, New Castle, DE, USA), specifically the 3220-TA Series III Model, equipped with a 5 mN load cell. To investigate the porous foams, glass plates were glued on the bottom and the cross head and the sample was placed between them. The measurement was performed first with a compression step with a strain rate of 0.015 mm s^{-1} up to 75% compression, followed by a release step with a strain rate of 0.015 mm s^{-1} down to 0% compression.

2.5. Electrochemical Characterization

The cell assembly was performed in an argon-filled glovebox (Labmaster SP, MBraun, Garching, Germany) and VWR connections made of PTFE with an inner diameter of 10.5 mm were used as cell housing. To avoid corrosion on the contacts, molybdenum rods were milled from pure molybdenum (99.95 at.% Mo) rods to until they had a diameter of 10.5 mm and fit tightly into the housing. Additionally, the contacts were wrapped with PTFE sealing tape, to exclude any influence of oxygen. The respective aerogel was placed on one side of the contact and six glass fiber separators (Grade 934-AH, Whatman, Merck KGaA, Darmstadt, Germany) with a diameter of 11 mm on top. 1-Ethyl-3-methylimidazolium chloride mixed with aluminum chloride in the ratio of 1:1.3 (IoLiTec, Ionic Liquid Technologies, Karlsruhe, Germany) was determined to be a suitable electrolyte. The separators and the aerogel were soaked with electrolyte (0.2 mL) and compressed using the other Mo contact as stamp. As counter electrode aluminum foil (99.99 at.% Al) was used. Galvanostatic charge/discharge tests were performed at a current density of 100, 1000 and 2000 mA g^{-1} in the voltage range of 0.2–2.0 V for the composite and 0.2 to 2.2 V for the rGO aerogel. Prior to testing, the cells sat for at least two hours to completely saturate the aerogel with electrolyte. Cyclic voltammetry was performed in a voltage window of 0.02–2.2 V, with a sweep rate of 0.1, 1, 10, 20, 50 and 100 mV s^{-1} for both aerogels. All measurements were performed on electrochemical test stations (VMP300, Biologic, Seyssinet-Pariset, France).

3. Results and Discussion

3.1. Fabrication of the Porous, Binder-Free rGO/SnO$_2$ Aerogels

Graphene oxide (GO) sheets were fabricated using a modified Hummers' method [34], resulting in sheet sizes ranging from 2 to 20 µm (Supporting Figure S1a). The SnO$_2$ nanoplatelets were synthesized by a hydrothermal approach [35]. Their dimensions were determined by scanning electron microscopy (SEM), revealing a thickness of a few nanometers and a lateral dimension of several tens of nanometers, with a tendency to form agglomerates with a size of 100 nm (Supporting Figure S1b).

The hybrid electrode was fabricated by adding SnO$_2$ nanoplatelets to an aqueous dispersion of the GO sheets. Soft sonication treatment was thereby performed to ensure a homogenous distribution (Figure 1) and exfoliation of GO sheets. Subsequently, the components were co-assembled by an ice-templating process. Ice crystals propagate thereby in a unidirectional manner along the applied temperature gradient in the z-direction, and thus restrict the solid load between them (Supporting Figure S2). Their subsequent sublimation, i.e., freeze-drying, leaves a highly porous, channel-like microstructure behind, in which the columnar pores are the replica of the removed ice crystals [36]. Such an ice-templating technique enables aligned, continuous channels throughout the whole height of the cylindrical aerogel. For the performance evaluation, aerogels composed solely of GO were fabricated, using the same concentration of GO in aqueous dispersion as that used for the hybrid electrode aerogel.

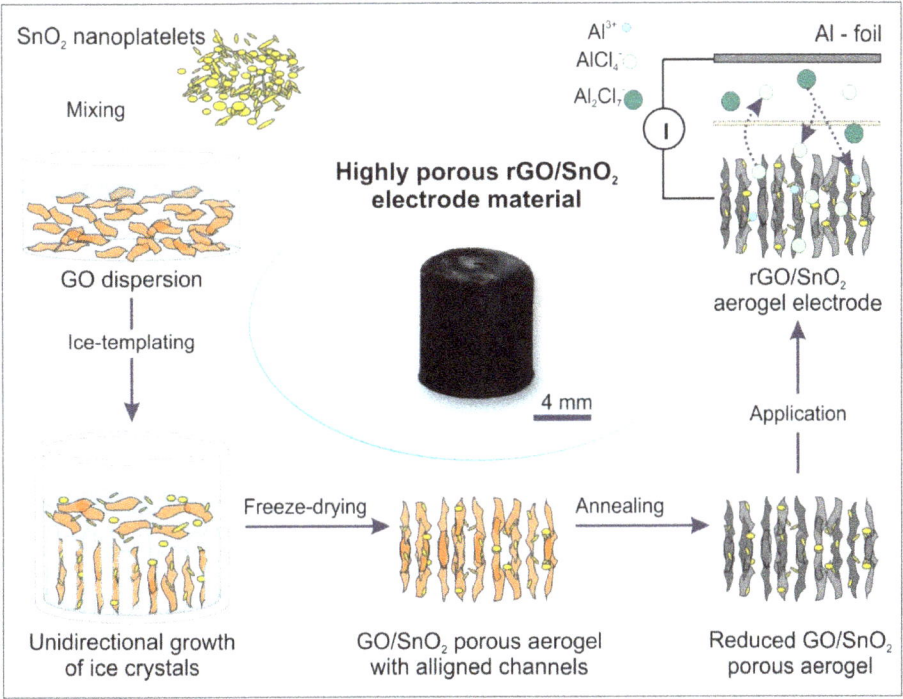

Figure 1. Schematic illustration of the fabrication process of free-standing highly porous rGO/SnO$_2$ aerogels and their application as a cathode material in AIBs.

The final step towards the fabrication of rGO/SnO$_2$ hybrid electrodes for AIBs (Figure 1) is annealing the aerogel to thermally reduce the GO. In this step, it is crucial to choose a slow heating rate to avoid fast gas evolution, which would lead to structural damage. To verify the structural preservation after annealing of the aerogels, pristine and hybrid, the microstructure was investigated by SEM in its original (Supporting Figure S3) and reduced state (Figure 2). To that end, a honeycomb-like pore structure is observed at the cross-section of the as-prepared and reduced aerogels (Figure 2a,d and Supporting Figure S3a,d), with a pore width in the range of 20 to 40 µm. From the longitudinal cross sections, the aligned pore channels are notable and a wall porosity in the range of tens of micrometers is revealed (Figure 2b,e and Supporting Figure S3b–e). The latter arises from local inhomogeneities of solid load between the ice crystals [37], coupled with the lateral size of the GO sheets, which is up to one magnitude smaller than the diameter of the ice crystals/channels. Moreover, a connection between the channels is observed in the longitudinal section (Figure 2c). The microstructural features identified for the aerogels are typical for ice-templated carbon materials [38–40]. Additionally, regarding the hybrid, agglomerates of SnO$_2$ nanoplatelets could be detected, which are wrapped between the GO sheets, asserting the co-assembly of both materials (Figure 2f and Supporting Figures S3f and S4). The striking similarity of the as-prepared and reduced aerogels' microstructures therefore ensured the minimal impact of the thermal treatment.

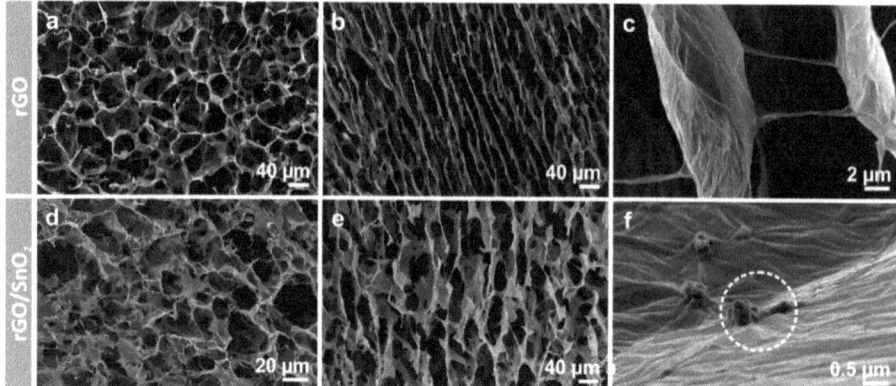

Figure 2. SEM images of the microstructures of rGO and rGO/SnO$_2$ aerogels (**a,d**) top view and (**b,e**) side view of the channels. Detailed view of the (**c**) GO walls and (**f**) the SnO$_2$ nanoplatelets embedded in the walls.

However, upon thermal treatment, a weight loss of 26% occurred, which is assigned to the removal of oxygen-containing functional groups and their reaction to gaseous CO and CO$_2$ [41]. A hybrid composition of 41 wt% rGO and 59 wt% SnO$_2$ is thereby obtained as deducted from thermogravimetric analysis (Supporting Figure S5). Moreover, highly conductive rGO is achieved due to the restoration of the sp^2-hybridization characteristic of graphene [42]. The removal of the oxygen-containing functional groups is further accompanied by a decrease in interlayer distance of neighboring sheets [43]. The powder X-ray diffraction (XRD) patterns of the hybrid aerogels, as-prepared and reduced, revealed a shift of the reflection at 11° for GO to 24° for rGO (Figure 3a,b). The reflection at around 11° originates from stacked GO, as it correlates with the pristine GO material. The shift therefore corresponds to a decrease in interlayer distance from 0.8 nm to 0.4 nm, which is close to the value of the interlayer distance in graphite (0.33 nm) [44]. This correlates to the results obtained by Raman, where an increase in the D- to G-band ratio is observed (Supporting Figure S6). Additionally, a peak broadening is observed for rGO, which correlates to a small crystallize size in the sheets [16]. Similarly, the same shift upon reduction is observed for the hybrid as a shoulder at 24° next to the reflection of SnO$_2$, indicating a decrease in layer distance of overlapping rGO sheets, which are located in the walls of the aerogel. However, the intensity of the shoulder is significantly lower and broader, compared to SnO$_2$, indicating a smaller diffracted crystal volume of rGO stacks as schematically presented in Figure 3b. This reduction in intensity is correlated with the decreased number of stacked rGO sheets, because SnO$_2$ particles are embedded between them hindering the restacking of rGO single sheets during removal of oxygen containing surface groups (see schematic in Figure 3b). XRD analysis shows that the SnO$_2$ nanoplatelets are nanocrystalline and exhibit a rutile structure, known as a good intercalation host [36,45–47]. After reduction, these rutile reflections of SnO$_2$ become more pronounced, indicating a crystal growth of the originally agglomerated nanostructured platelets into larger crystals (Supporting Figure S4). The crystallite size of the SnO$_2$ nanostructures was calculated using the Scherrer equation as 6.33 nm.

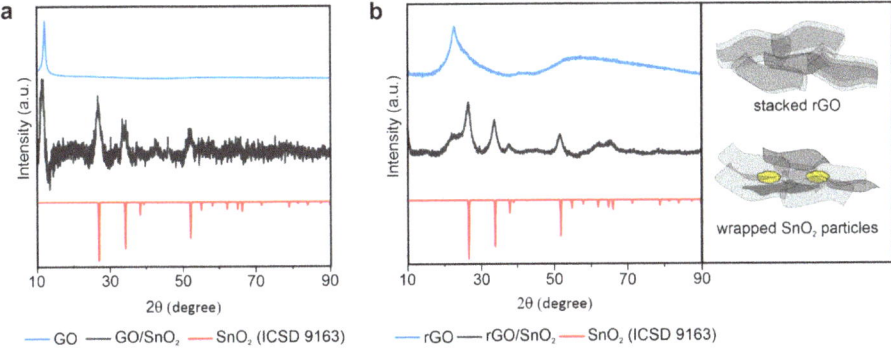

Figure 3. XRD pattern of the (**a**) GO/SnO$_2$ and (**b**) rGO/SnO$_2$ hybrid aerogel with a schematic presentation of the rGO sheets arrangement with and without SnO$_2$ particles.

Brunauer–Emmett–Teller (BET) measurements were performed on the pristine as well as the hybrid aerogels to determine their surface area and mesoporosity, as these features are known to highly influence the electrochemical performance of the active material [16]. A large surface area of 221.3 m^2 g^{-1}, in the case of the hybrid rGO/SnO$_2$, was thereby determined, 60% of that of the pristine rGO aerogel (367.9 m^2 g^{-1}). The presence of mesoporosity was also verified by the obtained N$_2$ adsorption–desorption isotherms, exhibiting a hysteresis form (Supporting Figure S7) [48]. This result is in good agreement with the presented microstructure (Figure 2b,e), revealing the location of the mesopores within the cell walls. However, the mesoporosity of the pristine aerogels is almost 2-fold that of the hybrid aerogels, as concluded from the cumulative pore volume of 0.47 cm^3 g^{-1} in the case of the pristine and 0.24 cm^3 g^{-1} in the case of the hybrid aerogel (Supporting Figure S7b). However, considering the significantly higher density of the hybrid aerogel (5.0 g cm^{-3}) an even larger cumulative pore volume fraction is achieved. Furthermore, the overall porosity $P = 1 - \rho_s/\rho$ *100% is estimated to 99.9% for both aerogels, pristine and hybrid, based on the density of carbon $\rho = 2.26$ g cm^{-3} and that of the aerogel ρ_s [44]. The scaffold structure, with its channel-like pores, is the major contributor to this high porosity. Notably, the tailored microstructure of the rGO/SnO$_2$ aerogels, combining the channel-like network of pores (macroporosity) with the cell wall mesoporosity, is promising. Specifically, the latter facilitates ion diffusion and enables the access to an increased number of intercalation and adsorption sites, while a homogenous distribution of the electrolyte throughout the electrode along the applied potential gradient is assured by the macroporosity [23].

3.2. Mechanical Performance of the rGO/SnO$_2$ Aerogels

The aerogels need to be compressed when implementing them in the battery cell as a free-standing cathode material. Mechanical testing of the aerogels was thereby conducted by means of compression, to ensure their structural integrity upon cell assembly and electrochemical analysis. The density and microstructure of the aerogels, specifically the degree of ordering and the alignment of the rGO sheets, thereby play a crucial role, defining the mechanical response. To this end, the size of the rGO sheets contributes immensely to the elastic mechanical performance. Sheets larger than 20 μm provide structural recovery upon deformation and a strength one order of magnitude higher than those smaller than 2 μm [38]. However, the size of the sheets also influences the electrochemical performance, whereas for smaller edge-rich sheets, an increased capacity in AIBs is obtained due to the additional intercalation sites at the edges [49]. Therefore, an optimal sheet size with sufficient mechanical stability and optimal electrochemical performance is crucial. The investigation of the rGO and rGO/SnO$_2$ aerogels (Supporting Figure S8), with an initial GO sheet size smaller than 2 μm, revealed a poor mechanical performance (Supporting Figure S9). Therefore, rGO and rGO/SnO$_2$ aerogels fabricated with a sheet size larger than 20 μm were investigated. Specifically, they would provide high elasticity and mechanical stability, similar to elastomeric materials, and yet sufficient edges for facilitated ion de-/intercalation. In particular, the continuous packing of the rGO within a cell wall and their interconnection promote their strength and superelasticity [38]. The shape of the obtained compressive stress–strain curves of the aerogels, whether pristine or hybrid (Figure 4a), indicate four typical deformation stages of cellular materials [50,51]. Specifically, stage I (strain < 5%) correlates to the elastic deformation, stage II (strain up to 50%) to cell wall buckling and stage III (strain up to 75%) to the densification of the aerogels [52]. Stage IV represents the recovery of the aerogels, referring to the flexibility of the structure with the rGO sheets. The macroscopic deformation of the aerogels is displayed in Figure 4b through digital images representative of the different stages (Supporting Movie S1 and S2). The elastic deformation of the aerogels is characterized by a Young's modulus similar for both aerogels, 0.41 ± 0.06 kPa for the pristine and 0.42 ± 0.09 kPa for the hybrid aerogel, with a slight increase in the compressive strength (Supporting Table S1). Considering that the presented aerogels exhibit a porosity of ~99.9%, these values are in good agreement to those obtained for other graphene/graphene oxide aerogels [52].

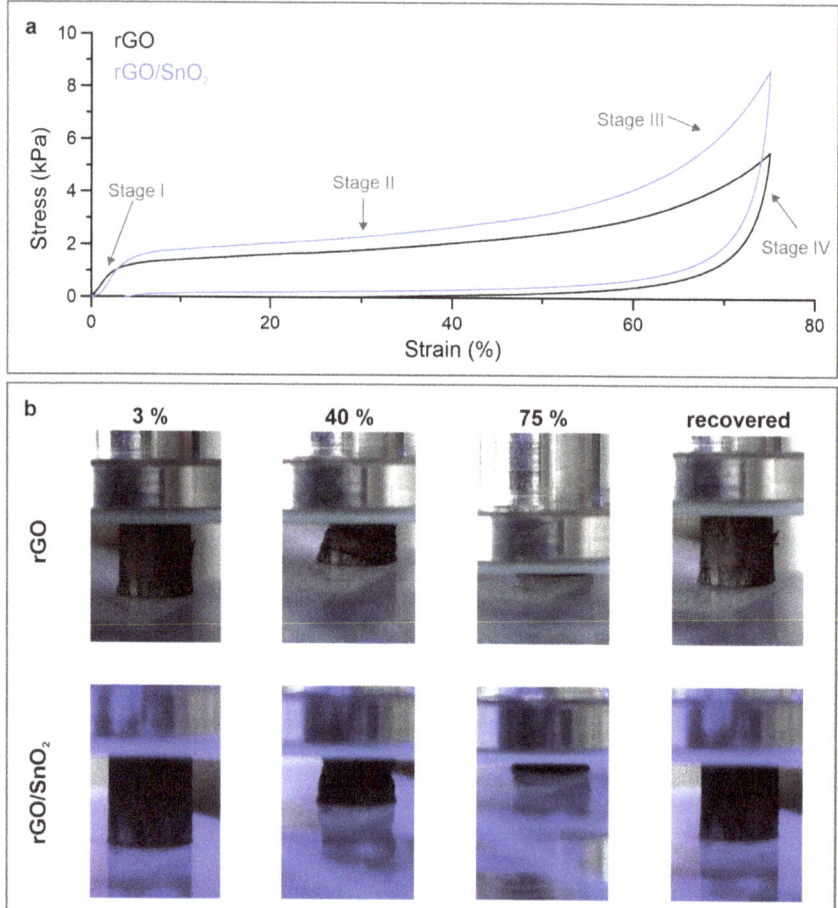

Figure 4. Mechanical performance of the rGO and rGO/SnO$_2$ aerogels under compression. (**a**) The compressive stress–strain curve with (**b**) the corresponding in situ images of the different compression states.

Notably, the rGO/SnO$_2$ aerogels show a superior recovery of 95.5% compared to pristine rGO aerogels (53.3%) (Figure 4a and Supporting Movie S3), allowing a reversible compression over a wide strain range with only a slight (3.5%) permanent deformation (Figure 4b). The latter is ascribed to defects in the aerogels' cell walls, such as micro-cracks [38], while the high recovery is attributed to the flexibility of the rGO sheets as well as the tailored structure-design, allowing significant energy absorption. The superior recovery of the hybrid aerogels is assumed to originate from the wrapping, or anchoring, of the SnO$_2$, which interconnects neighboring rGO sheets and interlock upon proceeding mechanical deformation. Such performance renders these hybrid aerogels as an ideal electrode material with a high mechanical stability, which maintains its flexibility and a certain porosity even under very high compression states. The accommodation of larger ions is thereby possible while the structural integrity of the electrodes is preserved.

3.3. Electrochemical Performance of the rGO and rGO/SnO$_2$ Aerogels

The ultra-light, free-standing and binder-free rGO and rGO/SnO$_2$ aerogels were electrochemically tested by directly soaking them with electrolyte and compressing them into discs upon cell assembly. The impact of additives used in conventional slurry-based electrodes on the electrochemical performance is thereby excluded, allowing to investigate the individual contribution of rGO and SnO$_2$ within the hybrid material. The electrodes were evaluated in the voltage window between 0.2 V and 2.2 V vs. Al/Al^{3+}. Figure 5a shows the corresponding cyclic voltammetry (CV) curves. For the pristine rGO aerogel, a plateau between 0.15 V and 1.7 V is observed [53]. This current–voltage behavior, where the current is predominantly linearly proportional to the voltage rate, is typical for non-Faradaic energy storage. In contrast, the hybrid rGO/SnO$_2$ electrode displays an anodic and a corresponding cathodic peak around 0.5 V (peak A), which is correlated with the de-/intercalation of Al^{3+} in SnO$_2$ [28]. Upon intercalation of the Al^{3+} into the rutile crystal structure of SnO$_2$, the tetra-valent Sn(IV) is reduced to divalent Sn(II), resulting in Al$_x$SnO$_2$ [28]. Whereas, the amount of intercalated Al^{3+} ions, denoted as x, reaches a maximum of 0.6. Interestingly, the CV curves of the hybrid electrode exhibits an additional anodic peak at 1.8 V and a corresponding cathodic peak at 1.3 V (peak B), which is not related to the intercalation into SnO$_2$. The origin of the peak is hypothesized to be the de-/intercalation of AlCl$_4^-$ at the interface between the SnO$_2$ particles and the rGO sheets.

This hypothesis is supported by results obtained for a freeze-dried rGO/SnS$_2$ composite [23]. They observed a peak with a large shift between positive and negative scan direction at a similar intercalation voltage.

Owing to the high surface area of the tailored aerogel electrodes, a large electrochemical double layer is formed. A significant capacitive contribution is thereby achieved analogous to the pristine rGO electrode (Figure 5a). The energy storage process is therefore a combination of de-/intercalation of the ions into SnO$_2$ as well as between rGO/SnO$_2$ and their adsorption on the rGO surface. Hence, an interplay between Faradaic and non-Faradaic processes occurs during the charging and discharging processes, using the active material to its utmost.

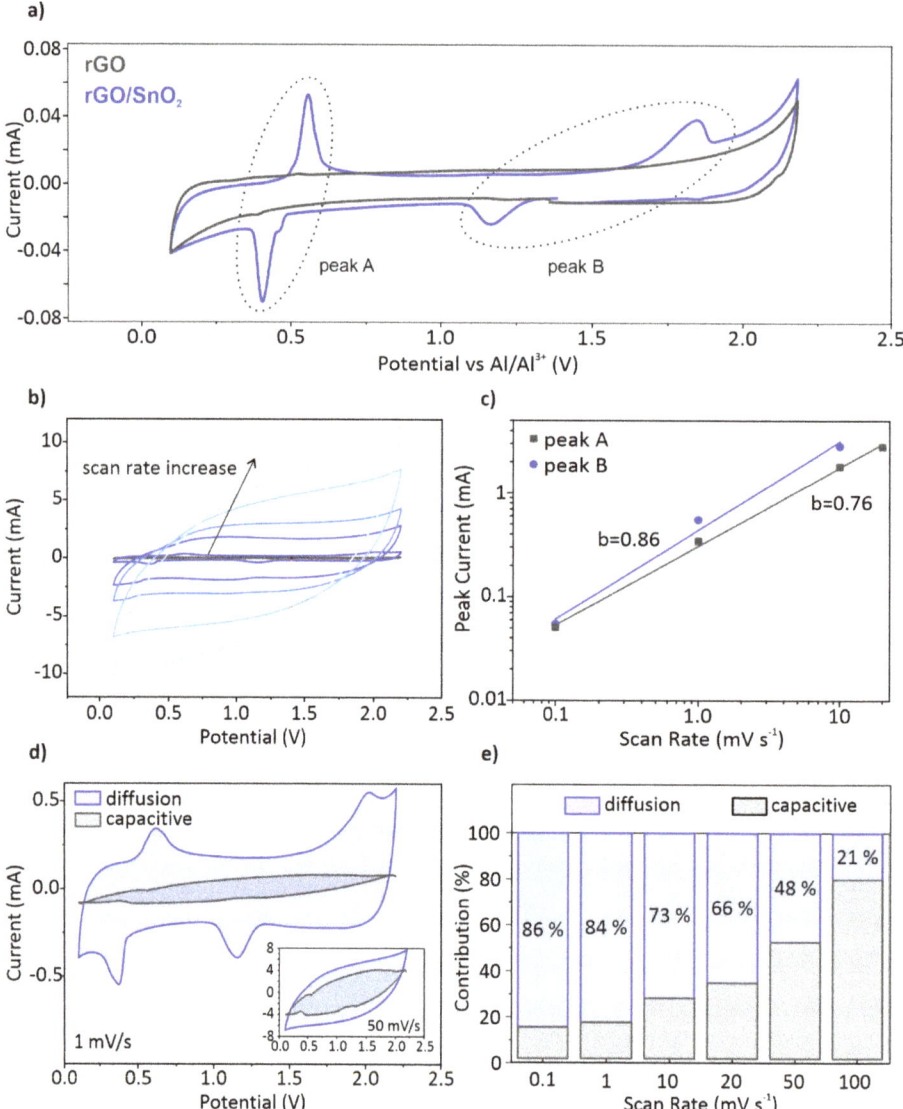

Figure 5. (**a**) Cyclic voltammograms (CV) of the rGO/SnO$_2$ hybrid electrode in comparison with the pristine rGO electrode at 0.1 mV s^{-1} (2nd cycle). (**b**) CVs at 0.1, 1, 10, 20, 50 and 100 mV s^{-1} for the rGO/SnO$_2$ hybrid electrode (2nd cycle). (**c**) Plot of the anodic peak current at the different scanning rates for b-value determination of peak A and peak B. (**d**) The contribution of the diffusion and capacitive controlled current response at 1 mV s^{-1} and 50 mV s^{-1} (inset). (**e**) The Faradaic and non-Faradaic contribution to the whole capacity (capacitive vs. diffusion-controlled energy storage) of the rGO/SnO$_2$ composite for all scanning rates, ranging from 0.1 mV s^{-1} up to 100 mV s^{-1}.

To distinguish the contribution of the diffusion-controlled processes to the overall energy storage from that of the pseudo-capacitive effect, CV measurements with scan rates ranging from 0.1 mV s^{-1} up to 100 mV s^{-1} were conducted (Figure 5b). At scan rates, up to 10 mV s^{-1}, the current peaks shift with a noticeable increase in peak separation. Upon increasing the scan rate to 100 mV s^{-1}, however, the peaks are not detectable anymore, and the energy storage process shifts to a mainly non-Faradaic characteristic. By plotting the peak current against the scan rate in a double-logarithmic plot (Figure 5c), indication of the predominant process can be shown [54]. The slope b of the thereby obtained linear regression follows $I = a\, v^b$, where I corresponds to the peak current, v to the scanning rate and a to the y-intersect. Values for b approaching unity are characteristic for capacitance-controlled processes, while values around 0.5 indicate diffusion-controlled processes [55]. The b-value is thereby determined as 0.76 for peak A and 0.86 for peak B, implying that both processes contribute. However, the higher value for peak B indicates a more capacitance-controlled process, denoting more surface-controlled kinetics. This additional peak arises therefore not only from intercalation of ions between SnO_2 and rGO, but mostly from their adsorption at the surface. To determine the contribution percentage of each process type, whether diffusion-controlled or surface-controlled, the corresponding current at different potential points is evaluated [18]. The cyclic voltammograms at 1 and 50 mV s^{-1} (Figure 5d) present the contribution of the different processes. A notable decrease of the capacitive current at the potential corresponding to the de-/intercalation processes, peak A and B, can be seen. Moreover, an increase of the capacitive contribution at the higher scan rate is apparent, comprising more than 50% at 50 mV s^{-1}. The contribution of the diffusion and capacitive processes for the various scanning rates is shown in Figure 5e. That of capacitive ion adsorption onto the surface of the electrodes increases with the charging rates, as anticipated. Surprisingly, the contribution of diffusion-controlled processes is dominant at a scanning rate of up to 50 mV s^{-1}. This behavior unites the main properties of a capacitive-controlled supercapacitor and a diffusion-controlled battery, combining high power and energy densities. Compared to a recently published work about graphite–graphite dual ion batteries using the same electrolyte, the diffusion-controlled contribution for the hybrid aerogel at the same scanning rate of 1 mV s^{-1} is more than two-fold increased [56]. This significantly increases the results from the synergy of nanosized SnO_2 particles wrapped by rGO sheets and the tailored microstructure of the hybrid aerogel electrode. Due to the small size of the SnO_2 particles, de-/intercalation is enabled at very high scanning rates of 100 mV s^{-1}, delivering a sizable contribution to the capacity (Figure 5e). The energy storage process is therefore optimized, simultaneously exploiting the large surface area of 221.3 m^2 g^{-1} induced by the tailored microstructure of the rGO-based aerogel electrode and the electrochemically active nanometer sized SnO_2 particles. The unidirectional channels, which originate from ice-templating, enable fast infiltration of electrolyte even at a compressed state. Additionally, they decrease the electron pathway and increase the high rate performance of the electrode [57]. This enables a supercapacitor-like performance at very high charging rates, coupled with comparatively large capacity values, thus overcoming the limitation in either energy or power density typically found in batteries or supercapacitors, respectively.

To further evaluate the rate performance of the hybrid aerogels in comparison to the pristine material, galvanostatic charge–discharge profiles with a charging rate of 2 C, 20 C and 40 C were conducted (Figure 6). A specific gravimetric capacity of the hybrid aerogel of 50 mAh g^{-1} was obtained for a current density of 100 mA g^{-1} (2 C). The impact of the SnO_2 nanoparticles addition on the electrochemical performance of the aerogels is concluded from the comparison of their volumetric capacities. The significant difference in gravimetric densities of the materials in question is thereby excluded.

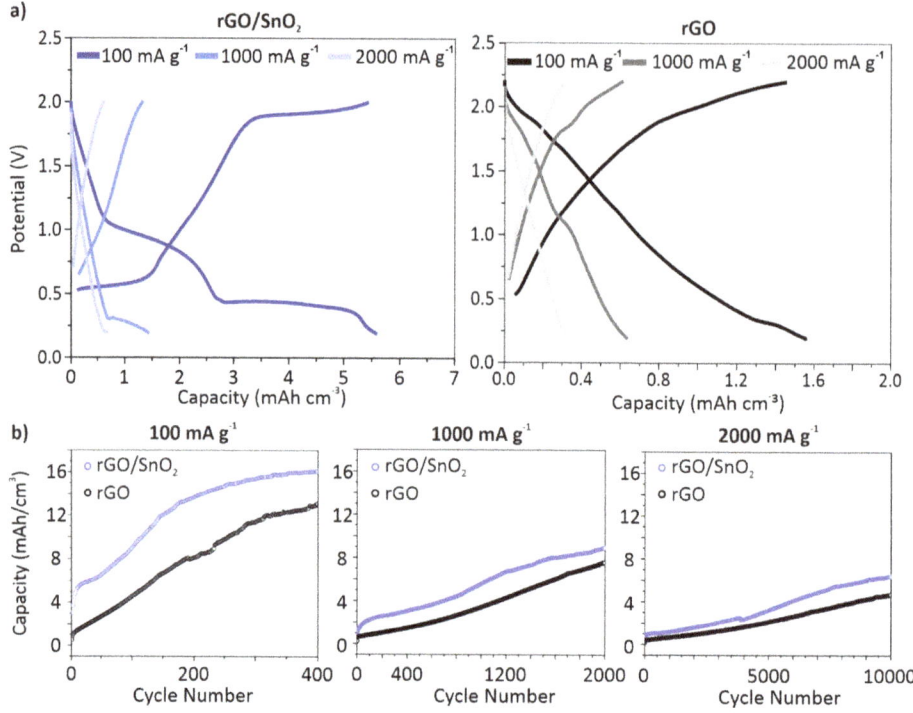

Figure 6. (**a**) Charge–discharge profiles of the hybrid rGO/SnO$_2$ (left) and pristine rGO (right) electrodes at the 10th cycle. (**b**) Rate performance of electrodes at a current density of 100, 1000 and 2000 mA g^{-1}. Their volumetric energy storage capacity was calculated for a compressed electrode with a thickness of 100 μm.

From the charge–discharge profiles (Figure 6a) of the hybrid electrode a large plateau around 0.5 V is observed, which correlates to peak A (Figure 5a). By increasing the charging rate, the length of the plateau decreases, equivalent to our findings from the CV analysis. However, the less pronounced plateau, corresponding to peak B (Figure 5a), disappears at higher charging rates. Additionally, a pseudocapacitive behavior is observed for the hybrid electrode, analogous to the pristine electrode. These findings corroborate the results obtained from the cyclic voltammetry analysis, demonstrating the synergy of the energy storage processes present in the hybrid electrodes. The volumetric capacity delivered by the hybrid electrode is accordingly higher by 23% than that of the pristine electrode. From the cell life investigation (Figure 6b), the same trend of the capacity evolution is observed for both electrodes. The increase of the capacities can therefore be attributed to the activation of the rGO and its subsequent reduction [41]. Furthermore, the initially high efficiency of 105% supports this conclusion (Supporting Figure S10). Moreover, by increasing the charging rate lower capacities are attained, reaching a capacity of 16.1 mAh cm^{-3} at 2 C and only 8.9 and 6.4 mAh cm^{-3} at 20 and 40 C, respectively. Notably, the hybrid electrodes deliver higher volumetric capacities than the pristine rGO electrodes, over 30% at the various charging rates. Compared to other cathode materials for AIBs the hybrid aerogel electrodes deliver adequate energy densities (20–50 Wh kg^{-1}) at notably high power densities (810–100 W kg^{-1}), as shown in the Ragone plot (Supporting Figure S11). On this basis, the benefit of the SnO$_2$ nanoparticle loading on the porous channel-like rGO aerogel is validated. Furthermore, the well-defined nanostructure of SnO$_2$ hinders the pulverization of the particles, leading to an excellent cycling stability over 10,000 cycles.

The electrochemical performance of the hybrid electrodes observed here can be attributed to three different energy storage mechanisms (Figure 7).

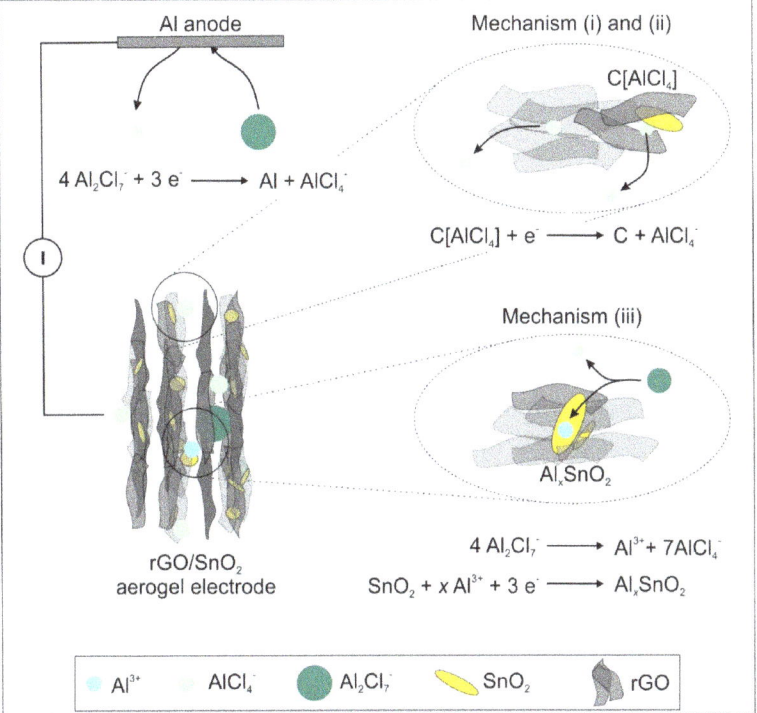

Figure 7. A schematic representation of the rGO/SnO$_2$ aerogel electrode in a half-cell configuration (left), where the three different energy storage mechanisms occurring at the cathode are presented with the corresponding simplified chemical reactions (right).

(i) The non-Faradaic contribution is partially correlated with the electrochemical double layer capacitance observed for rGO sheets with crystallite sizes below 10 nm [16]. The chloroaluminate anions are thereby adsorbed at the surface of rGO sheets.

(ii) Peak B is hypothesized to correlate with the intercalation of AlCl$_4^-$ between SnO$_2$ and rGO. The SnO$_2$ nanoparticles embedded in the rGO aerogel, provide sufficient space for intercalation, as they increase the cumulative pore volume fraction. When considering graphitic materials, intercalation involving AlCl$_4^-$ results in a characteristic peak around 2 V [13,57]. Additionally, the SnO$_2$ nanocrystals distort the graphitic structure, which could facilitate the intercalation and thus lowers the voltage. This mechanism contributes further to the pseudocapacitive behavior of the hybrid aerogel electrode.

(iii) The Faradaic contribution entails the de-/intercalation of Al^{3+} into SnO$_2$ (peak A), which was similarly observed for SnO$_2$/C cathodes [28].

4. Conclusions

We fabricated a highly porous, free-standing rGO/SnO$_2$ aerogel as a binder-free cathode for AIBs. The tailored microstructure of these aerogels provides a synergy of mechanical stability and enhanced electrochemical performance. It is characterized by aligned channels, the walls of which comprise rGO sheets and embedded SnO$_2$ nanoparticles, resulting in high flexibility with a significant structural recovery upon mechanical compression, up to 95.5%. Furthermore, the here-achieved integration of nanosized SnO$_2$ particles into the rGO aerogel creates a synergistic effect, where pseudocapacitive and diffusion-controlled energy storage mechanisms simultaneously contribute to the deliverable capacity and prolong the battery life, over 10,000 cycles. An enhancement of the volumetric capacity by 23% is thereby achieved, as opposed to its rGO aerogel counterpart. In addition, a notably high power density of 810 W kg^{-1} was achieved for the hybrid aerogels, comparable or superior to that of state-of-the art hybrid electrodes for AIBs. A step towards solving the current predicament in mobile energy devices is thereby made, as a combination of high power and high energy density, achievable by a single energy storage device, is attained. Our findings therefore provide new insights into the conceptual design of high-performance hybrid electrodes with pseudocapacitive behavior for AIBs.

Supplementary Materials: The following are available online at http://www.mdpi.com/2079-4991/10/10/2024/s1, Figure S1: SEM images of (**a**) a GO sheet, prepared by the modified Hummers Method (size between 2 to 20 µm). (**b**) SnO$_2$ nanoplatelets, prepared by a hydrothermal approach; Figure S2: Schematic illustration of the (**a**) ice-templating setup and (**b**) the evolution of unidirectional ice crystal growth, Figure S3: SEM images of the microstructures of GO and GO/SnO$_2$ aerogels (**a**,**d**) top view and (**b**,**e**) side view of the channels. Detailed view of the (**c**) GO walls and (**f**) the SnO$_2$ nanoplatelets embedded in the walls, Figure S4: Highly magnified SEM images of the (**a**) GO/SnO$_2$ and (**b**) rGO/SnO$_2$ hybrid aerogel. In both cases, the sheets wrap the SnO$_2$ nanoplatelets and hold them in place, Figure S5: Thermogravimetric analysis of the annealing process at 500 °C in argon and its corresponding mass loss of 26% and the determination of the rGO to SnO$_2$ ratio in synthetic air at 1100 °C, showing that 41 wt% of the composite is comprised of rGO, whereas 59 wt% is comprised of SnO$_2$, Figure S6: Raman Spectra of GO/SnO$_2$ (grey) and rGO/SnO$_2$ (red) samples, where the D- and G-bands are visible, Figure S7: Isotherm obtained by N2 adsorption and desorption curves of rGO and rGO/SnO$_2$ aerogels and their respective cumulative pore volume, Figure S8: SEM images of the microstructures of rGO and rGO/SnO$_2$ aerogels, which were fabricated using a GO dispersion with much smaller sheet sizes. (**a**,**d**) Top view and (**b**,**e**) side view of the channels. Detailed view of the (**c**) GO walls and (**f**) the SnO$_2$ nanoplatelets embedded in the walls, Figure S9: Mechanical performance of the rGO and rGO/SnO$_2$ aerogels, which were fabricated using a GO dispersion with smaller sheet sizes (<2 µm), under compression. (**a**) The compressive stress-strain curve with (**b**) the corresponding in-situ images of the different compression states, Figure S10: Efficiency plot of both the pristine and the hybrid aerogel, Figure S11: Ragone plot of the hybrid rGO/SnO$_2$ aerogel in comparison to the most recent literature in the field of AIB's. Supplementary Movie 1. High magnification movie of the compression test on a rGO/SnO$_2$ scaffold. Supplementary Movie 2. High magnification movie of the compression test on a rGO scaffold. Supplementary Movie 3. Practical demonstration of the scaffolds flexibility. Supplementary Table S1. Mechanical properties of rGO and rGO/SnO$_2$ aerogels. Supplementary Table S2: Mechanical properties of rGO and rGO/SnO$_2$ aerogels, which were fabricated using a GO dispersion with much smaller sheet sizes.

Author Contributions: Conceptualization, T.J., Z.B., L.R.; methodology, T.J.; investigation, T.J., D.H.; writing—original draft preparation, T.J.; writing—review and editing T.J., A.K., L.R., Z.B., A.M.D.; supervision, Z.B., J.B.; All authors have read and agreed to the published version of the manuscript.

Funding: This research received no external funding.

Acknowledgments: We highly appreciate the financial and technical support from Prof. J. Spatz and his group from the Max Planck Institute for Medical Research, Gerd Maier performed XRD measurements, whereas Erik Farley and Wenzel Gassner supported with SEM measurements. Maximilian Hackner assisted in TGA investigations and Traugott Wörner aided in the construction of the ice-templating setup. The BET measurements were performed by Sebastian Emmerling from the Department of B. Lotsch at Max Planck Institute for Solid State Research in Stuttgart.

Conflicts of Interest: The authors declare no conflict of interest

References

1. Lu, Y.; Zhang, Q.; Li, L.; Niu, Z.; Chen, J. Design Strategies toward Enhancing the Performance of Organic Electrode Materials in Metal-Ion Batteries. *Chemistry* **2018**, *4*, 2786. [CrossRef]
2. Luo, W.; Shen, F.; Bommier, C.; Zhu, H.; Ji, X.; Hu, L.N. Na-Ion Battery Anodes: Materials and Electrochemistry. *Acc. Chem. Res.* **2016**, *49*, 231–240. [CrossRef] [PubMed]
3. Valma, C.; Buchholz, D.; Passerini, S. Non-aqueous potassium-ion batteries: A review. *Curr. Opin. Electrochem.* **2018**, *9*, 41–48. [CrossRef]
4. Li, M.; Lu, J.; Ji, X.; Li, Y.; Shao, Y.; Chen, Z.; Zhong, C.; Amine, K. Design strategies for nonaqueous multivalent-ion and monovalent-ion battery anodes. *Nat. Rev. Mater.* **2020**, *5*, 276–294. [CrossRef]

5. Cui, L.; Zhou, L.; Kang, Y.-M.; An, Q. Recent Advances in the Rational Design and Synthesis of Two-Dimensional Materials for Multivalent Ion Batteries. *ChemSusChem* **2020**, *13*, 1071–1092. [CrossRef]
6. Das, S.K.; Mahapatra, S.; Lahan, H. Aluminium-ion batteries: Developments and challenges. *J. Mater. Chem. A* **2017**, *5*, 6347. [CrossRef]
7. Das, S.K. Graphene: A Cathode Material of Choice for Aluminum-Ion Batteries. *Angew. Chem. Int. Ed.* **2018**, *57*, 16606. [CrossRef]
8. Elia, G.A.; Marquardt, K.; Hoeppner, K.; Fantini, S.; Lin, R.; Knipping, E.; Peters, W.; Drillet, J.-F.; Passerini, S.; Hahn, R. An Overview and Future Perspectives of Aluminum Batteries. *Adv. Mater.* **2016**, *28*, 7564. [CrossRef]
9. Leisegang, T.; Meutzner, F.; Zschornak, M.; Münchgesang, W.; Schmid, R.; Nestler, T.; Eremin, R.A.; Kabanov, A.A.; Blatov, V.A.; Meyer, D.C. The Aluminum-Ion Battery: A Sustainable and Seminal Concept? *Front. Chem.* **2019**, *7*, 268. [CrossRef]
10. Yang, H.; Li, H.; Li, J.; Sun, Z.; He, K.; Cheng, H.-M.; Li, F. Die wiederaufladbare Aluminiumbatterie: Möglichkeiten und Herausforderungen. *Angew. Chem.* **2019**, *131*, 12104. [CrossRef]
11. Armand, M.; Endres, F.; MacFarlane, D.R.; Ohno, H.; Scrosati, B. Ionic-liquid materials for the electrochemical challenges of the future. *Nat. Mater.* **2009**, *8*, 621. [CrossRef] [PubMed]
12. Zhang, X.; Jiao, S.; Tu, J.; Song, W.-L.; Xiao, X.; Li, S.; Wang, M.; Lei, H.; Tian, D.; Chen, H.; et al. Rechargeable ultrahigh-capacity tellurium–aluminum batteries. *Energy Env. Sci.* **2019**, *12*, 1918. [CrossRef]
13. Lin, M.-C.; Gong, M.; Lu, B.; Wu, Y.; Wang, D.-Y.; Guan, M.; Angell, M.; Chen, C.; Yang, J.; Hwang, B.-J.; et al. An ultrafast rechargeable aluminium-ion battery. *Nature* **2015**, *520*, 324. [CrossRef]
14. Wu, Y.; Yi, N.; Huang, L.; Zhang, T.; Fang, S.; Chang, H.; Li, N.; Oh, J.; Lee, J.A.; Kozlov, M.; et al. Three-dimensionally bonded spongy graphene material with super compressive elasticity and near-zero Poisson's ratio. *Nat. Commun.* **2015**, *6*, 1. [CrossRef] [PubMed]
15. Chen, H.; Guo, F.; Liu, Y.; Huang, T.; Zheng, B.; Ananth, N.; Xu, Z.; Gao, W.; Gao, C. A Defect-Free Principle for Advanced Graphene Cathode of Aluminum-Ion Battery. *Adv. Mater.* **2017**, *29*, 1605958. [CrossRef] [PubMed]
16. Smajic, J.; Alazmi, A.; Batra, N.; Palanisamy, T.; Anjum, D.H.; Costa, P.M.F.J. Electrochemical Energy Storage: Mesoporous Reduced Graphene Oxide as a High Capacity Cathode for Aluminum Batteries. *Small* **2018**, *14*, 1870251. [CrossRef]
17. Wu, Y.; Gong, M.; Lin, M.-C.; Yuan, C.; Angell, M.; Huang, L.; Wang, D.-Y.; Zhang, X.; Yang, J.; Hwang, B.-J.; et al. 3D Graphitic Foams Derived from Chloroaluminate Anion Intercalation for Ultrafast Aluminum-Ion Battery. *Adv. Mater.* **2016**, *28*, 9218. [CrossRef]
18. Augustyn, V.; Simon, P.; Dunn, B. Pseudocapacitive oxide materials for high-rate electrochemical energy storage. *Energy Environ. Sci.* **2014**, *7*, 1597. [CrossRef]
19. Zhang, K.; Lee, T.H.; Cha, J.H.; Jang, H.W.; Shokouhimehr, M.; Choi, J.-W. S@ GO as a High-Performance Cathode Material for Rechargeable Aluminum-Ion Batteries. *Electron. Mater. Lett.* **2019**, *15*, 720. [CrossRef]
20. Gao, T.; Li, X.; Wang, X.; Hu, J.; Han, F.; Fan, X.; Suo, L.; Pearse, A.J.; Lee, S.B.; Rubloff, G.W.; et al. A Rechargeable Al/S Battery with an Ionic-Liquid Electrolyte. *Angew. Chem. Int. Ed.* **2016**, *55*, 9898. [CrossRef]
21. Wang, S.; Jiao, S.; Wang, J.; Chen, H.-S.; Tian, D.; Lei, H.; Fang, D.-N. High-Performance Aluminum-Ion Battery with CuS@C Microsphere Composite Cathode. *ACS Nano* **2017**, *11*, 469. [CrossRef] [PubMed]
22. Zhang, X.; Wang, S.; Tu, J.; Zhang, G.; Li, S.; Tian, D.; Jiao, S. Flower-like Vanadium Suflide/Reduced Graphene Oxide Composite: An Energy Storage Material for Aluminum-Ion Batteries. *ChemSusChem* **2018**, *11*, 709. [CrossRef] [PubMed]
23. Hu, Y.; Luo, B.; Ye, D.; Zhu, X.; Lyu, M.; Wang, L. An Innovative Freeze-Dried Reduced Graphene Oxide Supported SnS2 Cathode Active Material for Aluminum-Ion Batteries. *Adv. Mater.* **2017**, *29*, 1606132. [CrossRef] [PubMed]
24. Wang, S.; Yu, Z.; Tu, J.; Wang, J.; Tian, D.; Liu, Y.; Jiao, S. A Novel Aluminum-Ion Battery: Al/AlCl3-[EMIm]Cl/Ni3S2@Graphene. *Adv. Energy Mater.* **2016**, *6*, 1600137. [CrossRef]
25. Cohn, G.; Ma, L.; Archer, L.A. A novel non-aqueous aluminum sulfur battery. *J. Power Sources* **2015**, *283*, 416. [CrossRef]
26. Zhang, X.; Zhang, G.; Wang, S.; Li, S.; Jiao, S.J. Porous CuO microsphere architectures as high-performance cathode materials for aluminum-ion batteries. *Mater. Chem. A* **2018**, *6*, 3084. [CrossRef]

27. Wang, H.; Bai, Y.; Chen, S.; Luo, X.; Wu, C.; Wu, F.; Lu, J.; Amine, K. Binder-Free V2O5 Cathode for Greener Rechargeable Aluminum Battery. *ACS Appl. Mater. Interfaces* **2015**, *7*, 80. [CrossRef]
28. Lu, H.; Wan, Y.; Wang, T.; Jin, R.; Ding, P.; Wang, R.; Wang, Y.; Teng, C.; Li, L.; Wang, X.; et al. A high performance SnO2/C nanocomposite cathode for aluminum-ion batteries. *J. Mater. Chem. A* **2019**, *7*, 7213. [CrossRef]
29. Yamazaki, T.; Mizutani, U.; Iwama, Y. Electrical Properties of SnO2 Polycrystalline Thin Films and Single Crystals Exposed to O2- and H2-Gases. *Jpn. J. Appl. Phys.* **1983**, *22*, 454. [CrossRef]
30. Fonstad, C.G.; Rediker, R.H. Electrical Properties of High-Quality Stannic Oxide Crystals. *J. Appl. Phys.* **1971**, *42*. [CrossRef]
31. Jahnke, T.; Knöller, A.; Kilper, S.; Rothenstein, D.; Widenmeyer, M.; Burghard, Z.; Bill, J. Coalescence in Hybrid Materials: The Key to High-Capacity Electrodes. *ACS Appl. Energy Mater.* **2018**, *1*, 7085–7092. [CrossRef]
32. Kwon, O.H.; Oh, J.H.; Gu, B.; Jo, M.S.; Oh, S.H.; Kang, Y.C.; Kim, J.-K.; Jeong, S.M.; Cho, J.S. Lithium Polymer Batteries: Porous SnO2/C Nanofiber Anodes and LiFePO4/C Nanofiber Cathodes with a Wrinkle Structure for Stretchable Lithium Polymer Batteries with High Electrochemical Performance. *Adv. Sci.* **2020**, *7*, 2070093. [CrossRef]
33. Cho, J.S.; Kang, Y.C. Nanofibers Comprising Yolk-Shell Sn@void@SnO/SnO$_2$ and Hollow SnO/SnO$_2$ and SnO$_2$ Nanospheres via the Kirkendall Diffusion Effect and Their Electrochemical Properties. *Small* **2015**, *11*, 4673–4681. [CrossRef] [PubMed]
34. Hummers, W.S., Jr.; Offeman, R.E. Preparation of Graphitic Oxide. *J. Am. Chem. Soc.* **1958**, *80*, 1339. [CrossRef]
35. Wang, C.; Du, G.; Ståhl, K.; Huang, H.; Zhong, Y.; Jiang, J.Z. Ultrathin SnO2 Nanosheets: Oriented Attachment Mechanism, Nonstoichiometric Defects, and Enhanced Lithium-Ion Battery Performances. *J. Phys. Chem. C* **2012**, *116*, 4000. [CrossRef]
36. Deville, S.; Saiz, E.; Nalla, R.K.; Tomsia, A.P. Freezing as a Path to Build Complex Composites. *Science* **2006**, *311*, 515–518. [CrossRef]
37. Deville, S. *Freezing Colloids: Observations, Principles, Control, and Use: Applications in Materials Science, Life Science, Earth Science, Food Science, and Engineering (Engineering Materials and Processes)*; Deville, S., Ed.; Springer International Publishing: Cham, Switzerland, 2017; pp. 351–438.
38. Ni, N.; Barg, S.; Garcia-Tunon, E.; Perez, F.M.; Miranda, M.; Lu, C.; Mattevi, C.; Saiz, E. Mesoscale assembly of chemically modified graphene into complex cellular networks. *Sci. Rep.* **2015**, *5*, 4328.
39. Zhu, C.; Han, T.Y.-J.; Duoss, E.B.; Golobic, A.M.; Kuntz, J.D.; Spadaccini, C.M.; Worsley, M.A. Highly compressible 3D periodic graphene aerogel microlattices. *Nat. Commun.* **2015**, *6*, 1–8. [CrossRef]
40. Nieto, A.; Boesl, B.; Agarwal, A. Multi-scale intrinsic deformation mechanisms of 3D graphene foam. *Carbon* **2015**, *85*, 299. [CrossRef]
41. Acik, M.; Lee, G.; Mattevi, C.; Pirkle, A.; Wallace, R.M.; Chhowalla, M.; Cho, K.; Chabal, Y.J. The Role of Oxygen during Thermal Reduction of Graphene Oxide Studied by Infrared Absorption Spectroscopy. *Phys. Chem. C* **2011**, *115*, 19761. [CrossRef]
42. Chen, Y.; Fu, K.; Zhu, S.; Luo, W.; Wang, Y.; Li, Y.; Hitz, E.; Yao, Y.; Dai, J.; Wan, J.; et al. Reduced Graphene Oxide Films with Ultrahigh Conductivity as Li-Ion Battery Current Collectors. *Nano Lett.* **2016**, *16*, 3616. [CrossRef] [PubMed]
43. Acik, M.; Chabal, Y.J. A Review on Reducing Graphene Oxide for Band Gap Engineering. *JMSR* **2012**, *2*, 101. [CrossRef]
44. Wiberg, N.; Holleman, A.F.; Wiberg, E.; Fischer, G. *Lehrbuch der Anorganischen Chemie; 102., Stark Umgearb. u. Verb.*; De Gruyter: Berlin, Germany; New York, NY, USA, 2007.
45. Zhou, X.; Wan, L.-W.; Guo, Y.-G. Binding SnO$_2$ Nanocrystals in Nitrogen-Doped Graphene Sheets as Anode Materials for Lithium-Ion Batteries—Zhou. *Adv. Mater.* **2013**, *25*, 2152–2157. [CrossRef] [PubMed]
46. Noerochim, L.; Wang, J.-Z.; Chou, S.-L.; Wexler, D.; Liu, H.-K. Free-standing single-walled carbon nanotube/SnO2 anode paper for flexible lithium-ion batteries. *Carbon* **2012**, *50*, 1289. [CrossRef]
47. Hwang, S.M.; Lim, Y.-G.; Kim, J.-G.; Heo, Y.-U.; Lim, J.H.; Yamauchi, Y.; Park, M.-S.; Kim, Y.-J.; Dou, S.X.; Kim, J.H. A case study on fibrous porous SnO2 anode for robust, high-capacity lithium-ion batteries. *Nano Energy* **2014**, *10*, 53. [CrossRef]
48. Schneider, P. Adsorption isotherms of microporous-mesoporous solids revisited. *Appl. Catal. A Gen.* **1995**, *129*, 157. [CrossRef]

49. Zhang, Q.; Wang, L.; Wang, J.; Xing, C.; Ge, J.; Fan, L.; Liu, Z.; Lu, X.; Wu, M.; Yu, X.; et al. Low-temperature synthesis of edge-rich graphene paper for high-performance aluminum batteries. *Energy Storage Mater.* **2018**, *15*, 361. [CrossRef]
50. Knöller, A.; Kilper, S.; Diem, A.M.; Widenmeyer, M.; Runčevski, T.; Dinnebier, R.E.; Bill, J.; Burghard, Z. Ultrahigh Damping Capacities in Lightweight Structural Materials. *Nano Lett.* **2018**, *18*, 2519. [CrossRef] [PubMed]
51. Ashby, M.F.; Medalist, R.F.M. The mechanical properties of cellular solids. *MTA* **1983**, *14*, 1755. [CrossRef]
52. Qiu, L.; Huang, B.; He, Z.; Wang, Y.; Tian, Z.; Liu, J.Z.; Wang, K.; Song, J.; Gengenbach, T.R.; Li, D. Extremely Low Density and Super-Compressible Graphene Cellular Materials. *Adv. Mater.* **2017**, *29*, 1701553. [CrossRef]
53. Jiao, H.; Wang, J.; Tu, J.; Lei, H.; Jiao, S. Aluminum-Ion Asymmetric Supercapacitor Incorporating Carbon Nanotubes and an Ionic Liquid Electrolyte: Al/AlCl3-[EMIm]Cl/CNTs. *Energy Technol.* **2016**, *4*, 1112. [CrossRef]
54. Raju, V.; Rains, J.; Gates, C.; Luo, W.; Wang, X.; Stickle, W.F.; Stucky, G.D.; Ji, X. Superior Cathode of Sodium-Ion Batteries: Orthorhombic V2O5 Nanoparticles Generated in Nanoporous Carbon by Ambient Hydrolysis Deposition. *Nano Lett.* **2014**, *14*, 4119. [CrossRef] [PubMed]
55. Chen, C.; Wen, Y.; Hu, X.; Ji, X.; Yan, M.; Mai, L.; Hu, P.; Shan, B.; Huang, Y. Na+ intercalation pseudocapacitance in graphene-coupled titanium oxide enabling ultra-fast sodium storage and long-term cycling. *Nat. Commun.* **2015**, *6*, 1–8. [CrossRef] [PubMed]
56. Li, Z.; Liu, J.; Niu, B.; Li, J.; Kang, F. A Novel Graphite–Graphite Dual Ion Battery Using an AlCl3–[EMIm]Cl Liquid Electrolyte. *Small* **2018**, *14*, 1800745. [CrossRef] [PubMed]
57. Zhao, Z.; Sun, M.; Chen, W.; Liu, Y.; Zhang, L.; Dongfang, N.; Ruan, Y.; Zhang, J.; Wang, P.; Dong, L.; et al. Sandwich Vertical-Channeled Thick Electrodes with High Rate and Cycle Performance. *Adv. Funct. Mater.* **2019**, *29*, 1809196. [CrossRef]

Publisher's Note: MDPI stays neutral with regard to jurisdictional claims in published maps and institutional affiliations.

 © 2020 by the authors. Licensee MDPI, Basel, Switzerland. This article is an open access article distributed under the terms and conditions of the Creative Commons Attribution (CC BY) license (http://creativecommons.org/licenses/by/4.0/).

Article

On the Formation of Black Silicon Features by Plasma-Less Etching of Silicon in Molecular Fluorine Gas

Bishal Kafle [1,*], Ahmed Ismail Ridoy [1], Eleni Miethig [1], Laurent Clochard [2], Edward Duffy [2], Marc Hofmann [1] and Jochen Rentsch [1]

1. Fraunhofer Institute for Solar Energy Systems (ISE), 79110 Freiburg im Breisgau, Germany; ahmed.ismail.ridoy@ise.fraunhofer.de (A.I.R.); eleni.miethig@gmail.com (E.M.); marc.hofmann@ise.fraunhofer.de (M.H.); jochen.rentsch@ise.fraunhofer.de (J.R.)
2. Nines Photovoltaics, Dublin 24, Ireland; l.clochard@nines-pv.com (L.C.); e.duffy@nines-pv.com (E.D.)
* Correspondence: bishal.kafle@ise.fraunhofer.de; Tel.: +49-761-4588-2183

Received: 6 October 2020; Accepted: 2 November 2020; Published: 6 November 2020

Abstract: In this paper, we study the plasma-less etching of crystalline silicon (c-Si) by F_2/N_2 gas mixture at moderately elevated temperatures. The etching is performed in an inline etching tool, which is specifically developed to lower costs for products needing a high volume manufacturing etching platform such as silicon photovoltaics. Specifically, the current study focuses on developing an effective front-side texturing process on Si(100) wafers. Statistical variation of the tool parameters is performed to achieve high etching rates and low surface reflection of the textured silicon surface. It is observed that the rate and anisotropy of the etching process are strongly defined by the interaction effects between process parameters such as substrate temperature, F_2 concentration, and process duration. The etching forms features of sub-micron dimensions on c-Si surface. By maintaining the anisotropic nature of etching, weighted surface reflection (R_w) as low as $R_w < 2\%$ in Si(100) is achievable. The lowering of R_w is mainly due to the formation of deep, density grade nanostructures, so-called black silicon, with lateral dimensions that are smaller to the major wavelength ranges of interest in silicon photovoltaics.

Keywords: dry etching; black silicon; photovoltaics

1. Introduction

The formation of high aspect ratio and/or large surface area sub-micron structures on silicon is of high interest for several applications, such as photovoltaics, micro-electro-mechanical systems (MEMS), photodetectors, and silicon anodes for lithium-ion batteries [1–4]. Meanwhile, in silicon photovoltaics, the formation of submicron features on c-Si surface has received increased attention in the field of Si photovoltaics due to its ability to dramatically reduce the surface reflection to a very low value so that the wafer turns "black" in appearance, a so-called black-silicon (B-Si). Application of such anti-reflective surfaces on single or monocrystalline silicon (mono-c-Si) and multicrystalline silicon (mc-Si) surfaces have shown an improved short-circuit current density (J_{SC}) of solar cells due to a higher absorption of incident light [1–3].

For applications requiring high volume manufacturing such as a photonic detector or a photovoltaic cell, reduction of process costs is inevitable. However, a high cost of ownership (COO) of the vacuum-based etching equipment might make its application in the photovoltaic industry difficult. Other wet-chemical texturing methods such as metal-catalysed chemical etching (MCCE) also promise low reflection on both mono c-Si and mc-Si wafers. However, this method has drawbacks, such as the need for multiple processing steps, the use of expensive process materials, cumbersome waste management, and a high likelihood of trace metal particles being present in the Si wafer.

As alternatives to wet-chemical etching processes, plasma-based activation of Fluorine-containing gases like F_2, XeF_2, SF_6, CF_4, and NF_3 was widely investigated in the past to perform etching of c-Si for different applications such as photovoltaics, micro-electro-mechanical systems (MEMS), optoelectronic devices, and optical filters [4,5]. Chemical dry etching promises to provide significant economic and technological advantages to both of the abovementioned processes. It is known that gases like F_2, XeF_2, and ClF_3 can etch Si spontaneously, even at room temperatures, without any need of plasma excitations [6–9]. Typically, a good selectivity towards various masking materials including metals, photo-resists, SiO_x, SiN_x, etc.; and an isotropic etching of Si with high etch rates is the most important criterion desired for plasma-less dry etchants in microelectronics and/or MEMS micromachining. In solar cell fabrication, however, the anisotropic nature of the surface roughness left on the Si surface after the etching process is the most important criterion for forming anti-reflective surfaces allowing higher light absorption [10].

In comparison to other spontaneously activated etchants like XeF_2 and ClF_3, F_2 is known to have a lower Si etch rate at room temperature [9]. However, thermal activation of F_2 gas at moderate temperatures can be used to etch Si wafers with reasonably high etching rates. It has been reported that F_2 leaves a rougher Si surface than XeF_2 after the plasma-less etching process at room temperature [6]. However, no detailed knowledge about the surface roughness left after the etching process exists in the literature. In this paper, we study the plasma-less etching of Si in F_2/N_2 gas mixture when the Si wafer is heated at moderate temperatures of up to 300 °C. In comparison to other studies on halogen etching of Si, the basis of this study is an industrially available etching tool for products that need a high volume manufacturing platform. We first provide a brief introduction to the experimental tool, followed by a detailed study about the influence of tool process parameters on etch rate, surface morphology, and the resulting surface reflection after the texturing process. The process parameter variation of the etching tool is performed by using design of experiments (DOE) and the resulting output data (etch rate and surface reflection) are analysed using statistical methods. Thereafter, the nucleation of etch pits and the evolution of surface roughness are investigated based on detailed microscopic observations, and activation energies are calculated for the F_2-Si reaction system. The etching process forms high aspect ratio B-Si features that have potential for different applications; we particularly focus on discussing the properties of nanostructures that qualify them to be used as anti-reflective layers in photovoltaics.

2. Materials and Methods

2.1. Experimental Tool

In this work, an atmospheric pressure dry etching (ADE) tool (Nines ADE-100) is used to carry out dry etching of monocrystalline silicon (mono c-Si) and multicrystalline silicon (mc-Si) wafers. The prototype etching tool is manufactured by Nines Photovoltaics and installed at Fraunhofer ISE to establish an etching process that can be easily adapted in high volume production. Figure 1 shows the schematic providing details of the external connections and the reactor of the ADE tool.

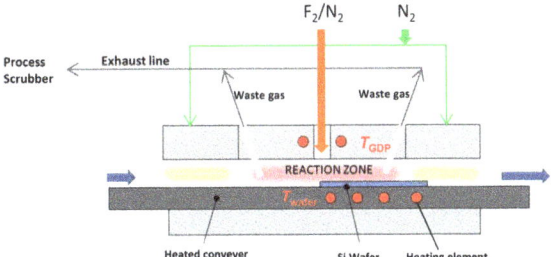

Figure 1. Schematics showing the atmospheric pressure dry etching (ADE) reactor and external connections to gas supply and exhaust lines.

All the gas lines are assembled in a valve manifold box (VMB) and mass flow controllers (MFC's) are used to control the gas flow rates. In the system, F_2 is the only etching gas present. High purity F_2/N_2 mixture in a gas bottle is stored in a gas cabinet. In this experiment, gas bottles filled with F_2/N_2 mixture with a maximum F_2 concentration of 20% are used. The etching is performed in the ADE tool, which is a compact and ventilated enclosure of metal sheets and polycarbonate doors. After the etching process, toxic and corrosive waste gases (F_2/SiF_x) are removed by the exhaust lines and fed to a dry bed process scrubber (CS clean systems) for abatement purposes. The etching gas is passed through a heated zone (gas diffusion plate GDP) intending to provide a temperature T_{GDP} that could potentially facilitate partial dissociation of F_2 into more reactive F atoms. N_2 is used as a carrier gas to dilute the effective F_2 concentration in the F_2/N_2 gas mixture during the etching process, as well as a purge gas to purge the gas lines and the reactor chamber after etching. Besides, the reactor is separated from the outer section of ADE tool with the help of two N_2 gas curtains, which are placed before and after the reaction zone. The gas curtains maintain a continuous flow of N_2 gas and contain the reactive gases inside the reactor. A slight pressure difference ($\Delta P \approx 60$ Pa) is maintained between the outside and inside of the reactor to contain the leakage of the reactive (both reactant and product) gases released during the etching process. The conveyer system is designed so as to transport the large area Si wafers (15.6 cm × 15.6 cm) through the reactor in an inline mode. The wafers are held in the conveyer system by a minor vacuum (2–3 kPa) and can be heated to a controlled temperature (T_{wafer}). The wafers are then dynamically transported through the reaction zone and later unloaded on the other side of the conveyer system.

The following nomenclatures of the tool process parameters are used in the paper: (a) flux of F_2 in F_2/N_2 gas mixture: Q_{F2}; (b) flux of separate N_2 as carrier gas: Q_{N2}; (c) total gas flux inside the reactor: Q_{F2+N2}; (d) effective concentration of F_2 in total gas flux: σ_F; (e) set temperature of the gas diffusion plate: T_{GDP}; (f) set temperature of the wafer substrate holder/heater conveyer: T_{wafer}; and velocity of the wafer substrate moving through the reaction chamber (v).

2.2. Design of Experiments

In order to perform the evaluation of the experimental results in the least biased way, the design of experiments (DOE) is performed using the statistical software. The major process parameters that might have an influence on the etching process are: T_{GDP}, T_{wafer}, v, Q_{F2+N2}, and σ_F. The total gas flux (Q_{F2+N2}) is kept constant, whereas the N_2 gas flux (Q_{N2}) is varied to reach the desired F_2 concentration (σ_F) values. Three level factorial design ($3 \times (k-p)$) with four factors ($k = 4$) and one block ($p = 1$) is used to generate the experimental design. Additionally, one replication is performed in each case, summing up the total number of experiments to be 27 × 2 = 54. Table 1 lists the process parameters as factors that influence the etching process, and Figure 2 shows the workflow of the experiment.

Table 1. Process parameters used for 3 × (k–p) experimental design.

Process Parameters	Units	Levels		
		Low	Medium	High
T_{GDP}	°C	200	245	290
T_{wafer}	°C	200	250	300
σ_F	%	1.67	3.33	5
v	mm/s	2	5	8

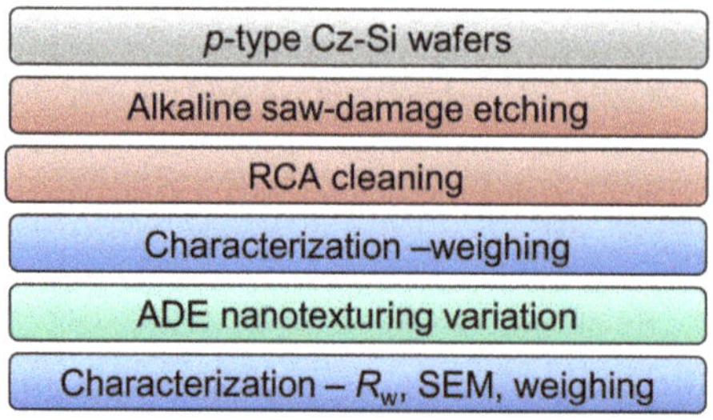

Figure 2. Process flow of the wafers used for the statistical design of experiments.

Large area (15.6 × 15.6 cm^2) p-type (100) Cz c-Si wafers are first saw-damage etched in alkaline solution and then cleaned using RCA sequence. The wafers are weighed in a weighing scale before transferring them to the etching tool. A variation of process parameters is performed in the etching tool as per the statistical design shown in Table 1. For each set of process parameters, the front side and the rear side of the wafer are etched consequently with the identical process parameters. In the analysis of the data, the etchings of front and rear side are assumed to be replicas of each other. After etching each side, weight measurements are performed to estimate the average value of the etch rate. Additionally, the surface reflectivity is measured in an integrated sphere using a UV/Vis/NIR spectrally resolved spectrophotometer (Varian Cary 5000, Agilent Technologies Germany GmbH, Waldbronn, Germany) in the wavelength spectrum of 250–1200 nm. The weighted surface reflection (R_w) [11] is then calculated in the wavelength spectrum of 300–1200 nm and the weighing function is applied using the internal quantum efficiency of a standard silicon solar cell under AM 1.5 G conditions.

2.3. Estimating Activation Energy

To investigate the influence of oxide termination on etching results, the Arrhenius behaviour of etching is investigated. The process plan of the experiment is shown in Figure 3.

The precursors used are p-type Cz wafers of (100) crystal orientation after saw-damage etching using alkaline solution. The wafers are then divided into three groups. All groups are cleaned separately using the cleaning sequence of hot HNO_3 (120 °C, 68%wt, 10 min), HF dip, and DI water rinsing. Afterwards, Group 2 is treated again with hot HNO_3 solution to grow a homogeneous chemical oxide on Si surface. Group 3 is kept in storage under the exposure of laboratory air for 4 days to grow a native oxide. Please note that the wet-chemical sequences for Group 1 and Group 2 are performed just before the etching process. The wafers from all three groups are etched together in the ADE tool at three different set T_{wafer} values. During etching, all other process parameters are kept constant. After the etching, the etch rate is calculated for each group.

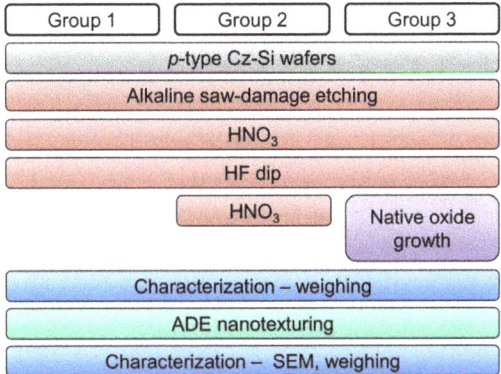

Figure 3. Process plan followed to investigate the Arrhenius behaviour.

The rate of the reaction (here, etch rate) can be expressed in the form of Arrhenius Equation (1)

$$R(Si) = k_0 e^{\left(\frac{-E_a}{RT}\right)}, \tag{1}$$

where $R(Si)$ represents the etch rate of Si in µm/min, k_0 is the pre-exponential factor in µm/min, R is the gas constant, and E_a is the activation energy in kCal/mol.

Based on above expression, $\ln(R(Si))$ is plotted against the inverse of T (T_{wafer}) and E_a is calculated.

2.4. Characterization of Nanostructures

p-type, (100) mono c-Si Cz wafers are first saw damage etched in alkaline solution, and cleaned by RCA cleaning followed by HF dip and DI water rinsing. The wafers are then etched using ADE process. During the ADE process, a variation of etching duration (v) is performed, whereas all other parameters are kept constant. Process conditions are chosen (T_{wafer} = 200 °C, Q_{F2+N2} = 24 slm, σ_F = 5.0%) in order to maintain directional etching in (100) direction, and the directionality of the etching process is verified by SEM measurements (SU 70, Hitachi High-Technologies Corporation, Tokyo, Japan). SEM top-view and cross-sectional view measurements of each sample are performed and five images of each sample are used for the analysis. The dimensions of the nanostructures are extracted by analysing SEM images by using image processing software ImageJ 1.48v [12]. The depth of the nanostructure is extracted simply as the distance from the top of the structure to its valley in the cross-sectional image. In Figure 8i, it was observed that the nanostructure top-view geometry resembles that of an ellipse. With this assumption, the 2D top-view image of the nanostructure is calculated by fitting it as an ellipse. From the measured areas, the diameter of a circular geometry is calculated for simplicity reasons, which represents the lateral dimension or width (w) of the nanostructure.

2.5. Estimation of Surface Enlargement Factor

The surface enlargement factors (S_f) of textured surfaces are estimated by two methods: (a) atomic force microscopy (AFM, Dimension 3100, Brucker Nano Surfaces (previously Veeco instruments & Digital Instruments, Santa Barbara, CA, USA) with a super sharp tip (Nanosensors SSS-NCH) with tip radius of 2 nm operated in the tapping mode; and (b) the change in weight of wafer after depositing 100 nm atomic layer deposited (ALD) Aluminium oxide (Al_2O_3) layer using a spatial ALD tool. The surface area increases with the amount of Si removed during the ADE process. The slope can be well fitted using a linear function (R^2 = 0.97) for the ALD deposition method and (R^2 = 0.93) for the AFM method, although a slight discrepancy in calculated S_f was observed. This is expected due to the inability of the AFM cantilever tip to reach the deep valleys of the nanotexture terrain, thus underestimating the S_f value.

3. Results

3.1. Statistical Variation of Process Parameters

The statistical analysis of the experimental section was performed for the etch rate and surface reflection as the dependent variables. It was observed that a normal distribution assumption of the residuals is valid and therefore analysis of variance (ANOVA) was used to analyse the output data.

3.1.1. Analysis of Etch Rate as the Dependent Variable

Half-normal plot was used to identify the statistically significant parameters influencing the etch rate and is shown in Figure 4.

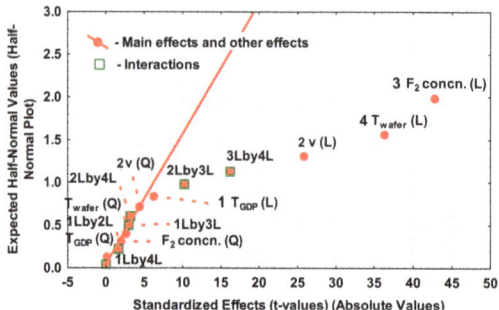

Figure 4. Half-normal plots to identify the significant factors to influence the etch rate. The straight lines in ii) are linear fits of the main effects (red filled circles) and interaction effects (green empty squares) that have an absolute value lower than 5. Here, 1, 2, 3, and 4 represent the process parameters T_{GDP}, v, σ_F, and T_{wafer} respectively. L and Q represent the type of effect (linear or quadratic) that the individual process parameters or their mutual interactions have on the dependent variable (here, etch rate). The mutual interaction effects between two parameters, for instance 1 and 2, are represented here as 1L by 2L and 1Q by 2Q for linear and quadratic interactions respectively.

This plot is based upon the assumption that all factors that have limited or no effect on responses (here etch rate) fall together, and their estimated effects (either main or interaction) can be fitted very well by a linear function. The outliers have higher statistical significance and the magnitude of the significance increases from left to right. Using this analysis, a large number of interaction effects can be discarded. The main effects of the σ_F, T_{wafer} and v are dominant in decreasing order. The linear interactions between σ_F-T_{wafer} and between σ_F-v are less significant. From the half-normal plot, it is observed that the temperature of the gas diffusion plate (T_{GDP}) is shown to have a very marginal effect on the etch rate. A relatively lower dissociation rate of F_2 is reported by Steudel et al. (degree of dissociation, $\alpha \approx 4\%$) at 1000 K [13], and ($\alpha < 1\%$) by Wicke et al. at 600 K [14]. The above reported measurements are performed at the chamber pressure of around 1 bar. Incidentally, this is close to the atmospheric pressure conditions that are used in our experimental set-up as well. Since almost no dissociation of F_2 is expected for the given experimental conditions, T_{GDP} is expected not to influence the etching process.

Apart from T_{GDP}, all other parameters are shown to linearly influence the etch rate in the experimental range of process parameters. Additionally, a linear interaction between σ_F-T_{wafer} and between σ_F-v is observed. These main and interaction effects can be intuitively understood by plotting the marginal mean and confidence intervals, as in Figure 5.

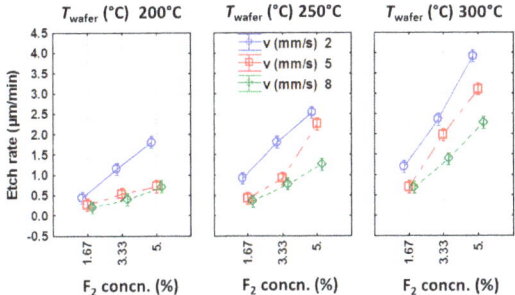

Figure 5. Plot of marginal means and confidence intervals showing interdependency of the most significant factors affecting etch rate. The lines are guides to the eye.

As expected, an almost linear increase in the etch rate is observed for an increasing σ_F irrespective of any values of v and T_wafer. A decrease in v always leads to a higher etch rate, which implies that the etch rate is increasing with the etching duration for each experiment. This is attributed to a possible increase in surface roughness and additionally to an increase in local temperature in the wafer due to the exothermic reaction between F_2 and Si. In the latter case, the subsequent heat release increases the reaction rate of the newly arriving F_2 molecules with Si. An increase in σ_F leads to a higher availability of F_2 molecules for the reaction with Si surface. This suggests that the etching process is still limited by the availability of F_2 in the reaction chamber within the range of process parameters applied in the experiment.

As per the rate equation, a higher temperature of silicon wafer is expected to enhance the etch rate due to an increment in the rate constant of the etching reaction. Here, the influence of increasing T_wafer on etch rate is marginal for process conditions featuring lowest F_2 concn. ($\sigma_F = 1.67\%$) and shortest process duration ($v = 8$ mm/s). For an increasing values of σ_F and v, the influence of T_wafer on the etch rate increases gradually. Meanwhile, for the combination of longest process duration ($v = 2$ mm/s) and highest fluorine concentration ($\sigma_F = 5\%$), T_wafer is found to strongly influence the etch rate. For instance, increasing $T_\text{wafer} = 200\ °C$ to $300\ °C$ resulted in a two-fold increment of the silicon etch rate at $\sigma_F = 5\%$ and $v = 2$ mm/s.

3.1.2. Analysis of R_w as the Dependent Variable

From the half-normal plots, v, σ_F, and T_wafer were identified as the main effects affecting R_w, whereas the interaction effects between v-T_wafer, and between v-σ_F are also dominant. In order to gain more insights about the main and interaction effects, marginal means and confidence intervals of the significant process parameters are plotted in Figure 6.

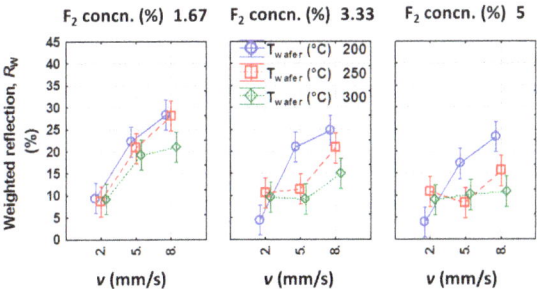

Figure 6. Plot of marginal means and confidence intervals showing interdependency of the most significant factors affecting R_w. The lines are guides to the eye.

Please note that for the lowest value of T_{wafer} (T_{wafer} = 200 °C), the graphs always show the same trend of an increasing R_w for an increase in v irrespective of the σ_F used during the etching process. For the T_{wafer} = 300 °C, R_w shows an increasing trend for an increase in v; however, only for the lowest σ_F = 1.67%. For the highest σ_F of 5%, the highest T_{wafer} = 300 °C leads to an almost constant R_w irrespective of the v used during the etching process. An optimum (lowest) value of R_w is achieved for the etching performed with a combination of the lowest T_{wafer} (200 °C), the highest σ_F (5%) and the lowest v (2 mm/s).

3.1.3. Change in Surface Morphology

A dramatic change in surface morphology is observed for the change in T_{wafer} if the etching is continued for the longest time period (v = 2 mm/s), and is summarized in Figure 7. At the lowest temperature (T_{wafer} = 200 °C), R_w gradually decreases for an increasing value of σ_F The representative cross-sectional SEM images indicate that anisotropic directional etching towards (100) direction occurs at this particular value of T_{wafer} for all values of σ_F, which results in the formation of conically shaped nanostructures in the c-Si surface.

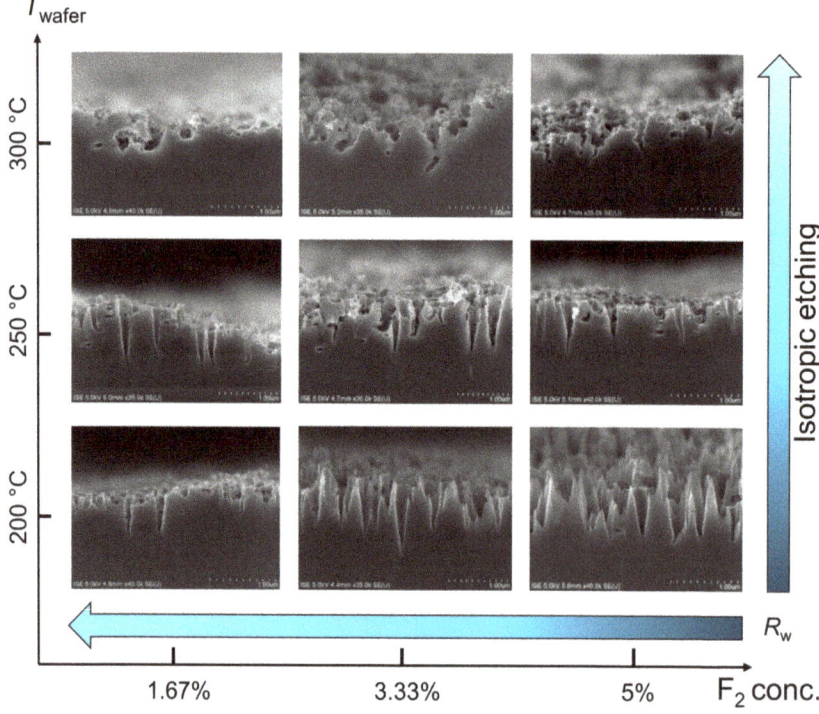

Figure 7. Cross-sectional SEM images of c-Si surfaces etched using constant velocity (v = 2 mm/s) using different combinations of F_2 concentration σ_F and temperature of the substrate T_{wafer}. The direction of the arrows represents an increasing value of R_w. The arrows represent an increasing value of R_w for decreasing σ_F, whereas an increasing degree of isotropic etching for an increasing T_{wafer}.

Here, a decrease in R_w for an increase in σ_F can be attributed to a higher density of nanostructures in the unit wafer area and to an increase in average depth of nanostructures that provides a higher grading of the refractive index from air to Si [15]. As the T_{wafer} increases to 250 °C, the directionality of the etching is disturbed and very shallow nanostructures start to form on top of the deeper cone-shaped nanostructures. At an even higher T_{wafer} = 300 °C, the deeper cone-shaped nanostructures almost disappear and the c-Si surface consists of only very shallow nanostructures, which, however, do not follow anisotropic etching in (100) direction anymore. The changes in surface structure can be clearly observed in the top-view SEM images of the etched surfaces, which are shown in Figure 8.

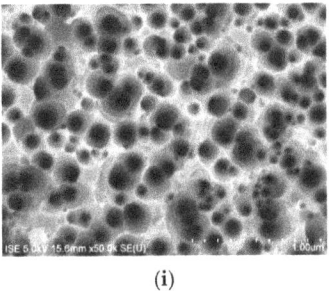 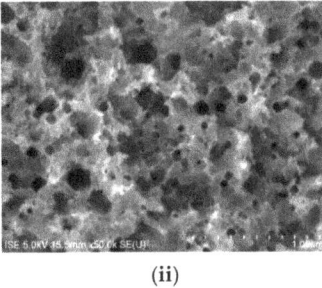

(i) (ii)

Figure 8. Top-view SEM images of c-Si surfaces etched with $\sigma_F = 5\%$ and $v = 2$ mm/s at T_{wafer} of (i) 200 °C, and (ii) 300 °C. The measured weighted surface reflection values are $R_w \approx 4\%$ ($T_{wafer} = 200$ °C) and $R_w \approx 9\%$ ($T_{wafer} = 300$ °C).

The absence of anisotropic cone-shaped nanostructures and the formation of very shallow nanostructures along various crystal planes of c-Si gradually increase the surface reflection. These very shallow nanostructures lead to a "sponge"-like appearance of the c-Si surface. Figures 7 and 8 suggest that an increase in T_{wafer} value is mainly dominating the change in surface morphology. However, it is observed that it is possible to compensate the effect of T_{wafer} by tailoring the values of v and σ_F. Cross-sectional SEM images of the surfaces that are etched at higher temperatures ($T_{wafer} = 250$ °C and $T_{wafer} = 300$ °C), which, however, still show etching in (100), are shown in Figure 9.

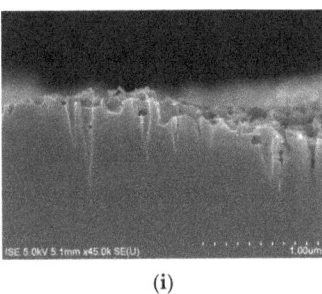 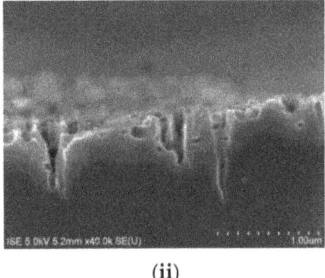

(i) (ii)

Figure 9. Cross-sectional SEM images showing possibilities of anisotropic etching at (i) $T_{wafer} = 250$ °C ($\sigma_F = 3.33\%$, $v = 5$ mm/s) and (ii) $T_{wafer} = 300$ °C ($\sigma_F = 5\%$, $v = 8$ mm/s). Both surfaces show an identical $R_w \approx 11\%$. The etch rates are measured to be 0.9 µm/min and 2.3 µm/min at $T_{wafer} = 250$ °C and $T_{wafer} = 300$ °C respectively.

These images provide a qualitative indication that the directional etching property can be maintained to a certain extent even at higher temperatures if F_2 availability and the duration of etching are controlled. This will be discussed in Section 4.

3.2. Initiation of F_2-Si Etching

Microscopic observations of F_2 etched c-Si surfaces, which were subjected to HF dip and DI-water rinse before performing the etching process, are used here to comment about the initiation of the F_2-Si etching process. Figure 10i presents the representative SEM images showing etch initiations in c-Si surfaces etched for different durations. It can be observed that there are three distinct areas in this image: (i) region R1 with no observable etching, (ii) region R2 with small etch pits, and (iii) region R3 with a more vigorous etching and slightly larger and deeper etch pits.

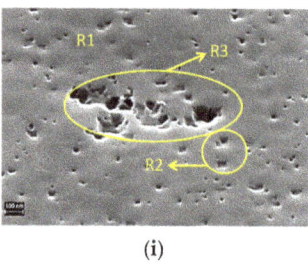
(i)

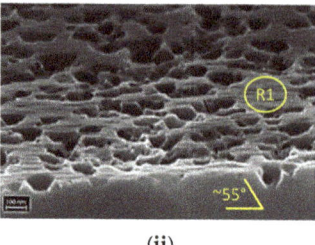
(ii)

Figure 10. SEM images showing etch initiations in freshly cleaned Si(100) surface after etching with a velocity of (i) 6.5 mm/s, and (ii) 3.5 mm/s.

From the first observations, the initial etching seems to start locally at certain locations that feature potentially higher local etch rates than others. Looking at R2 and R3, it becomes obvious that either the nucleation of pits and/or the very initial phases of their propagation show an anisotropic nature in Si(100). The preferential onset of etching for certain locations could be related to the formation of non-homogeneous native oxide during the waiting time between DI-water rinse and F_2-Si etching process. Although native oxide is reported to have negligible growth until at least 100 min after performing DI-water rinsing [16], heating of the Si wafer with $T_{wafer} > 170$ °C could accelerate the native oxide formation. The abundant pinholes in the oxide layer could provide reaction sites to start the etching reaction. Besides the presence of oxide species, vacancies, defects, and atomic steps are typically known to have a widespread presence in cleaned Si(100) surfaces [17]. Meanwhile, a local increase in roughness of the Si surface during the preparation of wafers for etching (RCA cleaning [18], HNO_3/HF based cleaning [19], saw-damage etching) can also promote etching by providing reaction sites. An account of F atoms adhering selectively at the reaction sites was reported previously for the HF solution treated Si(100) surface [19].

The anisotropic behaviour of the initial etching becomes more pronounced in the Figure 10ii as the inverted pyramid-like structures are clearly distinguishable. Additionally, a characteristic angle of ≈55 degrees between (100) and (111) crystal planes is observed that indicates that the initial F_2-Si etching is anisotropic in nature. This is expected for the F_2-Si etching system because of its sole chemical nature. Anisotropic etching is a known phenomenon typically observed during the etching of Si by alkaline solutions such as KOH, NaOH, TMAH, etc., and is due to the lowest density of surface atoms in (100) among all crystal planes. The side-walls evolve in (111) plane, which is the slowest etching plane due to a much higher density of Si-Si atoms.

3.3. Influence of Surface Termination on Activation Energy

Table 2 compares the activation energies of the F_2-Si etching process measured in current investigations to the ones that are previously reported by other authors for F/F_2 based etching of Si. It is observed that the surface reaction between F_2 and Si shows Arrhenius behaviour with a negative slope for an increasing temperature in all cases. This underlines the fact that the F_2-Si etching reaction is strongly dependent on surface temperature, and suggests that the reaction rate is limited by surface reaction kinetics. It is observed that E_a is lowest for the freshly cleaned wafer, slightly increases for the Si surface with native oxide, whereas it is almost twice as high when chemical oxide is grown. Meanwhile, E_a calculated for the freshly cleaned Si(100) wafer in this experiment ($E_a = 12.90 \pm 0.13$ kCal/mol) is found to be almost 40% higher than the ones reported by Mucha et al. [20] and Chen et al. [21], which is justified by the use of high vacuum in their etching apparatus. Furthermore, F atoms reportedly have a significantly lower E_a [22]. Meanwhile, one should be extremely cautious to conclude the influence of temperature on the reaction mechanism just based on these "apparent" E_a values. This is because the formation and the decomposition of SiF_x layer is reported to be temperature dependent and their properties also govern the etch rate [23].

Table 2. Comparison of E_a of F_2-Si etching estimated in this section to those measured by other authors in various conditions. No previous account of estimation of E_a is available for F_2-Si etching in atmospheric pressure (Atm.) conditions.

Etching Species	Starting Surface	E_a (kCal/mol)	Reactor Pressure	Reference
F_2	Si (100) after HF dip	12.9 ± 0.1	Atmospheric	This work
F_2	Si (100) with native oxide	13.7	Atmospheric	This work
F_2	Si (100) with HNO_3 oxide	23.0	Atmospheric	This work
F_2	Si (100), Si(110), Si (111)	8.0	Vacuum	[16]
F_2	Si (100)	9.3 ± 1.8	Vacuum	[17]
F	Si (100)	2.5 ± 0.1	Vacuum	[18]

4. Discussion

4.1. Etching Mechanism

It is observed that the main effects of σ_F, T_{wafer}, and v mainly determine the etch rate. An increasing etch rate is obtained for a decreasing v, an increasing T_{wafer}, and an increasing σ_F. In addition, interaction effects are found to be marginally significant. The mutual interaction of parameters can be understood using simple schematic in Figure 11, which shows the dependency of the reaction rate on different process parameters. An increase in etch rate for a decreasing v is expected to be a cumulative effect of a subsequent increase in surface roughness, and an increase in surface temperature ΔT_v due to the exothermic reaction between F_2 and Si. An increase in σ_F increases the reaction rate as per the rate equation. Simultaneously, it also increases the T_{wafer} due to the additional heat released $\Delta T_{concn.}$ as a result of an increased etch rate. A higher value of T_{wafer} increases the rate constant (k), and thereby the etch rate.

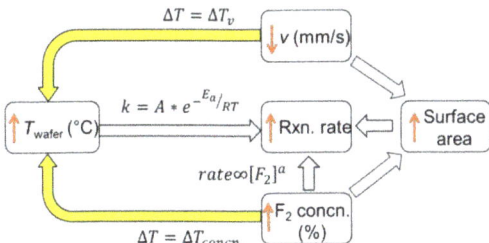

Figure 11. Schematics showing the influence of increasing concentration of the reactant (F_2) and increasing process duration (lowering v) on the resulting temperature and reaction rate of the exothermic reaction system we are analysing here. Here, the temperature dependence of the rate constant (k) is shown by the Arrhenius equation with A as pre-exponential factor, E_a as activation energy, and R as the ideal gas constant.

At a particular time t after the onset of the chemical reaction, for the case that the set temperature of the wafer substrate holder (T_{wafer}) is kept constant but σ_F and the etching duration is increased, the effective local temperature of the Si wafer (T_{Si}) can be defined as:

$$T_{Si} = T_{wafer} + \Delta T_v + \Delta T_{concn.,} \qquad (2)$$

where T_{wafer} represents the initial set temperature of the wafer, and $\Delta T_v + \Delta T_{concn.}$ represents the increase in wafer temperature due to the heat released depending on the duration of the etching process that also featured an increase in σ_F. Based upon experimental observations, it can be asserted that the absolute value of T_{wafer} is much higher than the factor $\Delta T_v + \Delta T_{concn.}$ within the experimental range of process parameters.

Based on the results and above discussion, a schematic model of the etching process is presented in Figure 12.

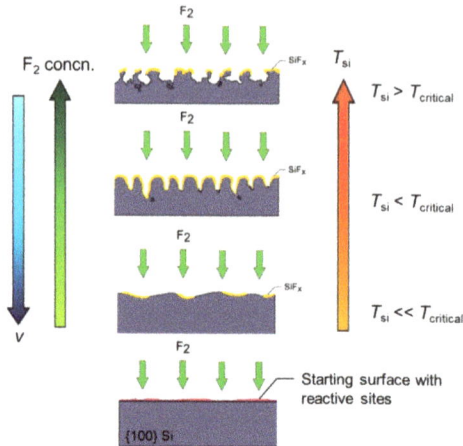

Figure 12. Schematic model of the etching process to explain anisotropic and isotropic etching of Si with F_2 depending upon the local Si surface temperatures (T_{Si}).

The F_2-Si reaction is expected to start initially at the reactive sites present in the starting c-Si surface, which is freshly cleaned (H-terminated). The reactive sites could be present due to (a) masking of Si by oxide islands, (b) inherent atomic-scale defects (defects, vacancies, steps) in the surface, and (c) evolution of very fine roughness from the preceding cleaning processes that included oxidizing agents. The differences in local etch rates lead to the nucleation of etch pits. Although the presence of native oxide islands definitely leads to a micro-masking and adds to the inhomogeneous etching behaviour of wafer locations in micron- and nano-scale, the preferential etching behaviour of F_2 already starts in the atomic scale and is proven by the STM measurements of Nakayama and Weaver [24]. Therefore, it is expected to be the major driving force in the nucleation of pits. It is proposed that the effective temperature of the wafer (T_{Si}) mainly determines the anisotropic nature of the etching in our experimental conditions. T_{Si} is a function of set wafer temperature (T_{wafer}) and the temperature increase (ΔT). The latter is the combination of the heat release during the F_2-Si etching process for the particular velocity (ΔT_v) and σ_F ($\Delta T_{concn.}$). According to Figure 11, the change in σ_F and v directly influences the resulting local temperature of the wafer. If the effective local temperature (T_{Si}) is less than a certain value, an anisotropic and directional etching of c-Si occurs. We call this value of T_{Si} as $T_{critical}$.

The initial etching is crystal-orientation dependent and leads to the formation of anisotropic features preferably in (100). An increase in surface temperature simultaneously increases the kinetic energy (K.E.) of the adsorbed F_2 molecules. This leads to an easier surface diffusion of the ad-atoms and allows them to relocate and bind to the reactive sites in the Si surface. This would lead to a faster etching. Furthermore, a higher surface temperature increases the fraction of molecules that have K.E. larger than the required activation energy to proceed with the reaction. This leads to higher etch rates in all crystal planes and an increase in the isotropic nature of etching. Additionally, the rate of formation of product species and its subsequent desorption from the Si surface also increases with an increase in surface temperature. An account of an increasing desorption probability of SiF_x species at higher temperatures is previously discussed by Winters and Coburn [23]. These product species are likely to behave as micro-masks on the Si wafer surface and their degradation with the temperature frees the reactive sites to the incident F_2 molecules. Hence, a more directional and anisotropic etching is to be expected at the lower effective temperature of Si ($T_{Si} < T_{critical}$), which leads to the formation of density grade nanostructures in (100) direction. The anisotropic etching mechanism holds true as long as the condition $T_{Si} < T_{critical}$ remains true, after which a competition between the anisotropic and the isotropic etching occurs. For $T_{Si} > T_{critical}$, isotropic etching is dominant and no deeper density grade structures are formed on the etched c-Si surface.

4.2. Nanostructure Properties

Obviously, for Si(100), process parameters should be chosen to maintain a directional etching, which allows formation of deep density graded nanostructures and lower R_W. By maintaining these conditions, it is observed in SEM investigations that the microscopic etch pits progresses into nanostructures with definite geometrical shapes. Figure 13 plots the surface enlargement factor S_f of the etched surfaces formed at different stages of etching. Here, process duration is varied to achieve various Si removal during etching, whereas all other process parameters (T_{wafer} = 200 °C, Q_{F2+N2} = 24 slm, σ_F = 5.0%) are kept constant. The process parameter combinations are chosen to ensure directional etching in (100), which is verified by using SEM investigations.

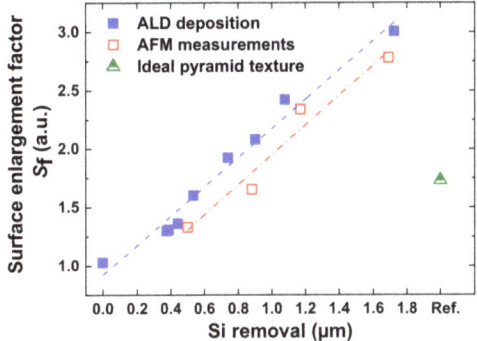

Figure 13. Surface enlargement factor (S_f) of textured surfaces estimated by using atomic layer deposited (ALD) deposition method and by atomic force microscopy (AFM) measurements. The dotted lines represent the linear fit to the data values. For comparison, S_f of an ideal alkaline (pyramid) texture is also plotted.

Here, S_f of the planar wafer (Si removal = 0 μm) wafer is measured to be 1.03. Meanwhile, S_f increases almost linearly with increasing Si removal during the etching process, leading S_f ~3.0 at 1.7 μm of Si removal. Figure 14 plots the extracted dimensions of the nanostructures formed after a different amount of Si removal.

An increasing removal of Si, the mean value of nanostructure depth (d_N) increases dramatically from 260 nm to up to 1822 nm. In the case of nanostructure width (w), an increase in the mean value of w is not clearly distinguishable due to a large standard deviation associated with the estimated data. Therefore, the influence of an increasing period of nanostructure on the surface reflection value is considered here as non-significant. Nevertheless, it should be noted that the maximum value of w is smaller than the wavelengths (λ_{light}) that are most important for Si photovoltaics (400–1000 nm). Under conditions of $w \leq \lambda_{light}$, the lowering of surface reflection occurs either due to the formation of effective medium ($w << \lambda_{light}$) and/or diffraction optics ($w \approx \lambda_{light}$).

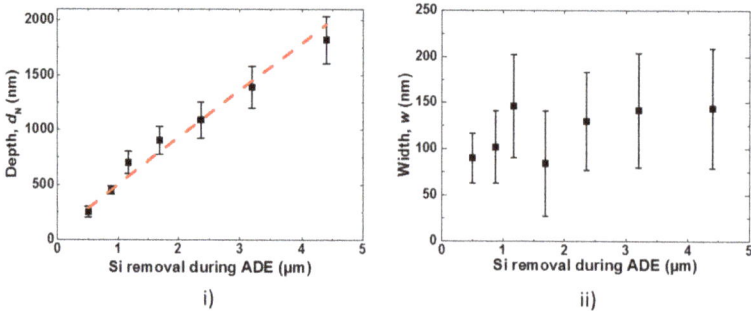

Figure 14. Plots showing change in (**i**) depth (d_N) and (**ii**) width (w) of nanostructures for an increasing amount of Si removal during ADE process.

In Figure 15i, measured weighted surface reflection (R_w) values are plotted against the estimated depth of nanostructures (d_N). R_w decreases gradually with an increasing value of d_N and the trend can be very well fitted by an exponential decay function ($R^2 = 0.98$). Meanwhile, the saturation of R_w occurs once the depth of the nanostructures exceeds a certain value. For instance, the weighted surface reflection value falls to $R_w \approx 5\%$ for $d_N \approx 700$ nm and to a low value of $R_w \approx 2\%$ for $d_N \approx 1100$ nm in the case of ADE nanotextured surfaces. In Figure 15ii, the normalized reflection value is plotted against the ratio of depth to wavelength (d_N/λ). The normalization of the reflection value (R_m) at each wavelength is performed with the measured reflection value of a saw damage-etched planar c-Si surface (R_0).

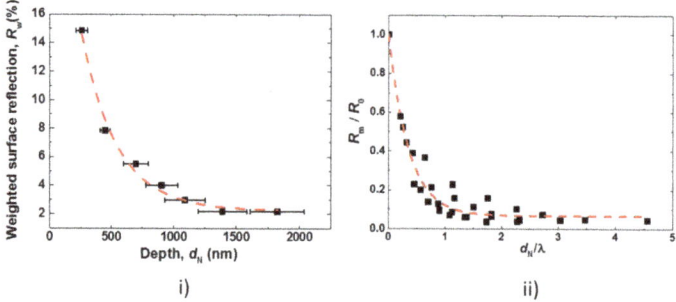

Figure 15. Plots showing (**i**) exponential reduction of normalized reflection for an increasing depth of nanostructures, and (**ii**) R_m/R_0 at each value of d_N/λ of nanotextured surfaces.

The influence of the scattering is minimized by not considering the wavelengths lower than 400 nm. Such a dimensionless quantity (d_N/λ) was previously used to explain the influence of nanostructures formed after MCCE on the surface reflection value [15]. The progression of nanostructure geometry is, however, significantly different for ADE texture compared to MCCE texture. Here, a scatter of the data points is observed, which is attributed to the possible systemic errors in extraction of nanostructure dimensions from SEM images. Nevertheless, the reflection R_m as a function of d_N/λ can be very well fitted by an exponential decay function ($R^2 = 0.97$).

$$R_m(d_N, \lambda) = R_0(d_N, \lambda)\left(A1 \times \exp\left(-\frac{d_N}{\lambda \times t1}\right) + y0\right) \quad (3)$$

The exponential decay fit to the above equation gives $y0 = 0.07$, $A1 = 0.93$, and $t1 = 0.35$. The plot suggests that the reflection of the nanostructured surface decreases to less than 10 times the value of the SDE surface if the depth of the nanostructure is comparable to the wavelength of interest, i.e., $d_N/\lambda \approx 1$. It was observed that in our case, the required values of grade depth to reach R_m/R_0 below 5% are higher ($d_N/\lambda \geq 2$) than the ones observed [15] for MCCE structures ($d_N/\lambda \geq 1$). This is attributed to a higher lateral dimension of ADE nanostructures in comparison to MCCE nanostructures.

5. Conclusions

In this paper, an alternative dry etching process is developed for its application in c-Si solar cells. The dry etching process utilizes spontaneous etching of Si by F_2 gas in atmospheric pressure conditions. The etching processes result in the formation of surface structures with dimensions in the sub-micron range, also known as nanostructures. Etching of Si by F_2 gas starts anisotropically and inverted pyramid-like structures are observed at the onset of etching. It is observed that the etching begins non-homogeneously in the Si surface. This phenomenon is attributed mainly to an accelerated attack of F_2 on surface defects and on the surface sites that are free from native oxide islands. It is proposed that the etching conditions result in an effective local surface temperature that is higher than a certain critical temperature, a highly isotropic etching of Si occurs and a "porous" looking Si surface is formed. In the other case, nanostructures with well-defined geometry and characteristic dimensions in sub-micron range are formed. Process parameters can be varied to reach even lower R_w values for an increasing Si removal during the etching process. This is correlated to an increase in the characteristic depths of the nanostructures, which dramatically lowers the weighted surface reflection (R_w) of c-Si in the wavelength spectrum of 400–1000 nm, the main range of interest for c-Si solar cells. As a consequence, a low value of weighted surface reflection $R_w \leq 2\%$ is achievable due to the formation of black silicon-like features.

Author Contributions: The following are the author contributions: Conceptualization, B.K. and M.H.; Data curation, B.K.; Formal analysis, B.K.; Funding acquisition, L.C., E.D., and J.R.; Investigation, B.K., A.I.R., and E.M.; Methodology, B.K., A.I.R., E.M., and M.H.; Project administration, L.C., E.D., M.H., and J.R.; Supervision, M.H. and J.R.; Validation, M.H.; Visualization, B.K.; Writing—original draft, B.K.; Writing—review & editing, L.C. and M.H. All authors have read and agreed to the published version of the manuscript.

Funding: This research was funded by European Union's Seventh Framework Programme managed by REA Research Executive Agency ([FP7/2007-2013] [FP7/2007-2011]) under grant agreement n° 286658.

Acknowledgments: The authors thank Jutta Zielonka for SEM measurements and Daniel Trogus for help in processing of the samples.

Conflicts of Interest: The authors declare no conflict of interest.

References

1. Fukui, K.; Inomata, Y.; Shirasawa, K. Surface Texturing Using Reactive Ion Etching For Multicrystalline Silicon Solar Cells. In Proceedings of the 26th PVSC, Anaheim, CA, USA, 29 September–3 October 1997.
2. Repo, P.; Haarahiltunen, A.; Sainiemi, L.; Yli-Koski, M.; Talvitie, H.; Schubert, M.C.; Savin, H. Effective Passivation of Black Silicon Surfaces by Atomic Layer Deposition. *IEEE J. Photovolt.* **2013**, *3*, 90–94. [CrossRef]
3. Otto, M.; Kroll, M.; Käsebier, T.; Salzer, R.; Tünnermann, A.; Wehrspohn, R.B. Extremely low surface recombination velocities in black silicon passivated by atomic layer deposition. *Appl. Phys. Lett.* **2012**, *100*, 191603. [CrossRef]
4. Jansen, H.; de Boer, M.; Legtenberg, R.; Elwenspoek, M. The black silicon method: A universal method for determining the parameter setting of a fluorine-based reactive ion etcher in deep silicon trench etching with profile control. *J. Micromech. Microeng.* **1995**, *5*, 115. [CrossRef]
5. Pezoldt, J.; Kups, T.; Stubenrauch, M.; Fischer, M. Black luminescent silicon. *Phys. Status Solidi* **2011**, *8*, 1021–1026. [CrossRef]
6. Arana, L.R.; Mas, N.d.; Schmidt, R.; Franz, A.J.; Schmidt, M.A.; Jensen, K.F. Isotropic etching of silicon in fluorine gas for MEMS micromachining. *J. Micromech. Microeng.* **2007**, *17*, 384–392. [CrossRef]
7. Saito, Y. Plasmaless etching of silicon using chlorine trifluoride. *J. Vac. Sci. Technol. B* **1991**, *9*, 2503. [CrossRef]
8. Winters, H.F.; Coburn, J.W. The etching of silicon with XeF2 vapor. *Appl. Phys. Lett.* **1979**, *34*, 70. [CrossRef]
9. Ibbotson, D.E.; Mucha, J.A.; Flamm, D.L.; Cook, J.M. Plasmaless dry etching of silicon with fluorine-containing compounds. *J. Appl. Phys.* **1984**, *56*, 2939. [CrossRef]

10. Kafle, B.; Seiffe, J.; Hofmann, M.; Clochard, L.; Duffy, E.; Rentsch, J. Nanostructuring of c-Si surface by F2-based atmospheric pressure dry texturing process. *Phys. Status Solidi A* **2015**, *212*, 307–311. [CrossRef]
11. Zhao, J.; Green, M.A. Optimized antireflection coatings for high-efficiency silicon solar cells. *IEEE Trans. Electron Devices* **1991**, *38*, 1925–1934. [CrossRef]
12. Hibbs, A.R. Imaging Software. In *Image Processing with ImageJ*; Abramoff, M.D., Magalhaes, P.J., Ram, S.J., Eds.; Kluwer Academic/Plenum Publishers: New York, NY, USA, 2004; pp. 163–176. ISBN 978-1-4757-0983-4.
13. Ralf Steudel. *Chemistry of the Non-Metals. Translation of Chemie der Nichtmetalle*; Walter de Gruyter & Co.: Berlin, Germany, 1976.
14. Wicke, E.; Franck, E.U. Physikalisch-chemische Eigenschaften des Fluors. *Angew. Chem.* **1954**, *66*. [CrossRef]
15. Branz, H.M.; Yost, V.E.; Ward, S.; Jones, K.M.; To, B.; Stradins, P. Nanostructured black silicon and the optical reflectance of graded-density surfaces. *Appl. Phys. Lett.* **2009**, *94*, 231121. [CrossRef]
16. Ohmi, T.; Isagawa, T.; Kogure, M.; Imaoka, T. Native Oxide Growth and Organic Impurity Removal on Si Surface with Ozone-Injected Ultrapure Water. *J. Electrochem. Soc.* **1993**, *140*, 804–810. [CrossRef]
17. Aldao, C.M.; Weaver, J.H. Halogen etching of Si via atomic-scale processes. *Prog. Surf. Sci.* **2001**, *68*, 189–230. [CrossRef]
18. Huff, H. *Into the Nano Era: Moore's Law Beyond Planar Silicon CMOS*; Springer: Berlin/Heidelberg, Germany, 2008; ISBN 9783540745594.
19. Deal, B.E.; Helms, C.R. (Eds.) *The Physics and Chemistry of SiO$_2$ and the Si-SiO$_2$ Interface*; Springer: New York, NY, USA, 1988.
20. Mucha, J.A.; Donnelly, V.M.; Flamm, D.L.; Webb, L.M. Chemiluminescence and the Reaction of Molecular Fluorine with Silicon. *J. Phys. Chem.* **1981**, *85*, 3529–3532. [CrossRef]
21. Chen, M.; Minkiewicz, V.J.; Lee, K. Etching Silicon with Fluorine Gas. *J. Electrochem. Soc.* **1979**, *126*, 1946. [CrossRef]
22. Flamm, D.L. The reaction of fluorine atoms with silicon. *J. Appl. Phys.* **1981**, *52*, 3633. [CrossRef]
23. Winters, H.F.; Coburn, J.W. Surface science aspects of etching reactions. *Surf. Sci. Rep.* **1992**, *14*, 162–269. [CrossRef]
24. Nakayama, K.S.; Weaver, J.H. Si (100) − (2 × 1) Etching with Fluorine: Planar Removal versus Three Dimensional Pitting. *Phys. Rev. Lett.* **1999**, *83*, 3210–3213. [CrossRef]

Publisher's Note: MDPI stays neutral with regard to jurisdictional claims in published maps and institutional affiliations.

© 2020 by the authors. Licensee MDPI, Basel, Switzerland. This article is an open access article distributed under the terms and conditions of the Creative Commons Attribution (CC BY) license (http://creativecommons.org/licenses/by/4.0/).

Article

Structural and Electrochemical Analysis of CIGS: Cr Crystalline Nanopowders and Thin Films Deposited onto ITO Substrates

Suzan Saber [1,2], Bernabé Marí [1,*], Andreu Andrio [3], Jorge Escorihuela [4,*], Nagwa Khattab [2], Ali Eid [2], Amany El Nahrawy [2], Mohamed Abo Aly [5] and Vicente Compañ [6,*]

1. Institut de Disseny i Fabricació, Universitat Politècnica de València, Camí de Vera s/n, 46022 Valencia, Spain; s.k.saber@hotmail.com
2. National Research Center, 33 El Bohouth St. (Former El Tahrir St.), Dokki, Giza, Cairo 12622, Egypt; nag_khb@yahoo.com (N.K.); suzankamal85@gmail.com (A.E.); amany_physics_1980@yahoo.com (A.E.N.)
3. Departament de Física Aplicada, Universitat Jaume I, Avda. Sos Baynat s/n, 12080 Castelló, Spain; andrio@uji.es
4. Departamento de Química Orgánica, Universitat de València, Avda. Vicent Andrés Estellés s/n, Burjassot, 46100 Valencia, Spain
5. Chemistry Department, Faculty of Science, Ain Shams University, Cairo 11566, Egypt; aboalymoh@hotmail.com
6. Departament de Termodinàmica Aplicada, Escola Tècnica Superior d'Enginyers Industrials (ETSII), Universitat Politècnica de València, Camí de Vera s/n, 46022 Valencia, Spain
* Correspondence: bmari@fis.upv.es (B.M.); jorge.escorihuela@uv.es (J.E.); vicommo@ter.upv.es (V.C.)

Citation: Saber, S.; Marí, B.; Andrio, A.; Escorihuela, J.; Khattab, N.; Eid, A.; Nahrawy, A.E.; Abo Aly, M.; Compañ, V. Structural and Electrochemical Analysis of CIGS: Cr Crystalline Nanopowders and Thin Films Deposited onto ITO Substrates. *Nanomaterials* 2021, 11, 1093. https://doi.org/10.3390/nano11051093

Academic Editor: Luca Pasquini

Received: 17 March 2021
Accepted: 20 April 2021
Published: 23 April 2021

Publisher's Note: MDPI stays neutral with regard to jurisdictional claims in published maps and institutional affiliations.

Copyright: © 2021 by the authors. Licensee MDPI, Basel, Switzerland. This article is an open access article distributed under the terms and conditions of the Creative Commons Attribution (CC BY) license (https://creativecommons.org/licenses/by/4.0/).

Abstract: A new approach for the synthesis of nanopowders and thin films of CuInGaSe$_2$ (CIGS) chalcopyrite material doped with different amounts of Cr is presented. The chalcopyrite material CuIn$_x$Ga$_{1-x}$Se$_2$ was doped using Cr to form a new doped chalcopyrite with the structure CuInxCr$_y$Ga$_{1-x-y}$Se$_2$, where x = 0.4 and y = 0.0, 0.1, 0.2, or 0.3. The electrical properties of CuIn$_x$Cr$_y$Ga$_{1-x-y}$Se$_2$ are highly dependent on the Cr content and results show these materials as promising dopants for the fabrication thin film solar cells. The CIGS nano-precursor powder was initially synthesized via an autoclave method, and then converted into thin films over transparent substrates. Both crystalline precursor powders and thin films deposited onto ITO substrates following a spin-coating process were subsequently characterized using XRD, SEM, HR-TEM, UV–visible and electrochemical impedance spectroscopy (EIS). EIS measurement was performed to evaluate the dc-conductivity of these novel materials as conductive films to be applied in solar cells.

Keywords: chalcopyrite compounds; nanocrystals; hydrothermal; spin coating; EIS; conductivity

1. Introduction

In the past decade, the photovoltaics (PV) technology has strongly evolved and even reached grid parity with other conventional energy sources [1]. In this regard, significant advances have been reported after long time of investigation in the field of thin film solar cells technology [2,3]. Although silicon has been the dominant material in the market of PV technology, others thin films options such as cadmium telluride (CdTe), copper indium gallium disulfide (CIGS2), and copper indium gallium diselenide (Cu(In,Ga)Se$_2$, CIGS) materials, which are capable of maintaining constant their efficiency for more than 15 years, have recently reached conversion efficiencies of 22.1% and 22.6%, respectively [4,5]. Furthermore, other new emerging technologies such hybrid perovskite solar cells have also improved efficiencies up to 22.1% in the past years [6–9]. This massive development is mainly dependent on the economic situation, as global prices have increased in the recent years in combination with the fuel sources depletion [10].

The first practical solar cell reported in the 1950s was mainly formed of crystalline silicon and with an efficiency around 4.5% [11]. Since then, a considerable increase in the materials used in the 1970s reached tens of absorbers. Afterwards nanotechnology and materials science have grown exponentially since the 1990s. This blossoming enriched the PV field and led to significant advances in this field. The most relevant output production of that era was the synthesis of CdTe and CIGS materials [12–15]. From the point of view of material properties, the solar cells are generally divided into two basic categories: bulk and thin-film solar cells [16,17]. The first group is made of monocrystalline silicon or polycrystalline silicon materials, whereas thin film photovoltaic cells are based on solution processable semiconductors. Materials such as CIGS and other materials from the same family have emerged as promising candidates to be used in thin-film solar cells due to their high absorption coefficient, changeable bandgap, and resistance to photo-degradation [18–21]. CIGS thin films are generally fabricated using non-vacuum techniques, such as co-evaporation and sputtering techniques [22]. However, vacuum methods have several drawbacks such as high cost, difficulty to produce large area coatings and low yields. One of the critical disadvantages facing CIGS development is the multistage processes such as co-evaporation and two-step processes involving sputtering followed by a selenization stage [17–28].

On the other hand, film deposition using non-vacuum techniques are at present the most widely used in industrial applications because they offer an apparent alternative to vacuum-based processes such as colloidal methods, sol–gel [29,30], paste coating [31], inkjet printing [32], and solvothermal processes [33,34]. From all of them, the solvothermal method constitutes the most competitive process, mainly due to the low cost and high efficiency, and actually has a great demand in the production of solar cells at the industrial scale [35]. Pre-deposition, co-evaporation incorporation, and post-deposition incorporation are the main strategies of deposition generally used for thin film preparation of CIGS chalcopyrite materials [36].

The doping process reported in the literature is mostly focused on alkali doping and some for-trace metal impurities incorporated from substrate material to absorption layers [37,38]. Alternative doping agents in the CIGS layer such as Fe impurities on the photovoltaic properties of the solar cells [39–41], and other metals such as Mn, V, Ti, Cr, Ni, and Al have also been evaluated on the performance of the solar cells [42,43]. However, a thorough study of how these impurities can act as defects, and influence on the electronic properties of the material is still scarce for the family of CIGS compounds.

Herein, we used a pre-deposition incorporation process form the preparation of CIGS compounds, which incorporates Cr in the precursor solution to afford the starting material of precursor powders. The thermal treatment performed was provided using autoclave (solvothermal) method. This precursor powders were used in in the preparation of thin films. In particular, we focused on the preparation of $CuIn_xCr_yGa_{1-x-y}Se_2$ with $x = 0.4$ and $y = (0.0, 0.1, 0.2, 0.3)$ with a well-controlled particle size in the order of nanometers. The crystalline precursor powders and thin films deposited onto indium tin oxide (ITO) glass substrates using a spin-coating process were subsequently characterized using X-ray diffraction (XRD), field emission scanning electron microscopy (FE-SEM), high-resolution transmission microscopy (HR-TEM), energy dispersive X-ray spectroscopy (EDX), UV–visible and electrochemical impedance spectroscopy (EIS). The analysis of results allowed us to quantify which is the doping percentage of Cr in the $CuIn_{0.4}Ga_{0.6}Se_2$ structure.

2. Materials and Methods

2.1. Synthesis of $CuIn_xCr_yGa_{1-x}Se_2$ Nano-Crystalline Precursor Powders

$CuInxGa_{1-x}Se_2$ was synthesized by the solvothermal method. For this, powders of Cu (99.9%), Se (99.99%), GaI_3 (99%), and $InCl_3$ (99.999%) from Sigma–Aldrich were used as copper, selenium, gallium, and indium sources, respectively. $Cr(ClO_4)_3$ was used as Cr^{3+} source and ethylenediamine (99.5%, Sigma–Aldrich Química SL, Madrid, Spain) was used as solvent. All compounds were introduced in a nitrogen-filled glove box and oxygen below 1 ppm. The amount of each precursors was 2 mmol of the copper elemental powder, 1 mmol of indium, gallium metal sources, and 4 mmol of selenium powder, respectively. Chromium percentage was calculated for every batch process in order to prepare different

Cr and Ga series at a fixed indium percentage to get this structure $CuIn_xCr_yGa_{1-x-y}Se_2$ with x = 0.4 and y = (0.0, 0.1, 0.2, 0.3).

All precursors were weighed and stirred overnight with a suitable amount of solvent. The stirred solution was then loaded into a Teflon autoclave container and annealed in an open-air oven for 36 h at 220 °C. The resultant product was washed several times using ethanol and distilled water to remove excess by-product or impurity materials. Centrifuging and ultra-sonication processes were repeated every washing time. The centrifuging process were conducted for 10 min at 4000 rpm. Finally, the completely purified precursor powders were dried at 100 °C for 2 h, and a black nano-crystalline powder of $CuIn_xCr_yGa_{1-x-y}Se_2$ was obtained.

2.2. Synthesis of the $CuIn_xCr_yGa_{1-x}Se_2$ Thin Films

All the collected synthesized precursor powders were washed several times with distilled water and ethanol in order to remove all sub product residue chemicals. These washed powders were then dried in open-air drier at 100 °C for 2 h. After this the powders were loaded into the glove box redispersion in the same preparation solvent as and left for stirring overnight then filtered using 0.2 μm mesh filters. Completely clear dissolved filtered precursors solution of the prepared nanocrystalline powders were obtained available for further processes.

Indium tin oxide (ITO) glass substrates were cut down into small pieces of the desired surface area. Cleaning process of the glass substrate were applied by washing the substrates using soap, water, ethanol, and acetone ultrasonically, dried, and then introduced into the glove box. The dissolved nanocrystalline precursor solution was then transformed to thin films by spin coating process. Metal salts nano-crystalline precursor powder to solvent ratio were taken in consideration (1:3). Supplementary Figure S1 depicts the setup use for the synthesis of $CuIn_xCr_yGa_{1-x}Se_2$ thin films. Subsequent thermal treatment using a pre-heated hot plate was applied for improving the crystal structure and reach the required film thickness.

2.3. Characterization of $CuIn_xGa_{1-x}Se_2$ Nano-Crystalline Powders and Thin Films

X-ray diffraction (XRD) was used to characterize the crystalline structure and phase of the prepared powder. The measurements were recorded in 2θ range from 20 to 65° with a Rigaku Ultima IV diffractometer in the Bragg–Bentano configuration using CuKα radiation (λ = 1.54060 Å). The particle size and morphology of the resultant CIGS precursor powder were characterized by field emission scanning electron microscopy (FESEM) of model Zeiss ULTRA 55 equipped with In-Lens and secondary electrons detectors. High-resolution transmission microscopy (HRTEM, 200 KV) JEOL Model: JEM-2100F (Tokyo, Japan) were used to study the crystalline structure and morphology of nanowire arrays. The chemical composition and the purity of the samples were characterized using energy dispersive X-ray spectroscopy (EDX). Finally, an Novocontrol broadband dielectric spectrometer (BDS) (Hundsangen, Germany) integrated by an SR 830 lock-in amplifier with an alpha dielectric interface was used for measuring the conductivity of the resultant precursor powders.

2.4. Electrochemical Impedance Spectroscopy (EIS) Measurements

The powders conductivity in the transversal direction were measured by impedance spectroscopy in the frequency interval of $10^{-1} < f < 10^7$ Hz applying a 0.1 V signal amplitude to ensure the linear response in a range of temperature between 20 and 200 °C. The measurements were carried out in dry and wet conditions using a Novocontrol broadband dielectric spectrometer (BDS) (Hundsangen, Germany) integrated by an SR 830 lock-in amplifier with an Alpha dielectric interface was used [44–46]. The powders were previously dried in a vacuum cell and their thicknesses were measured afterwards using a micrometer, taking the average of 10 measurements in different parts of the surface. Next the samples were sandwiched between two gold circular electrodes coupled to the impedance spectrometer acting as blocking electrodes. Before to start the measurement the assembly powder-electrode was annealed in the Novocontrol setup under an inert dry nitrogen atmosphere. For this, firstly a temperature cycle from 20 to 200 °C and then from 200 to 20 °C, in steps of 20 °C, was carried out. After this, in a new cycle of temperature scan, the dielectric spectra were collected in each step from 20 to 200 °C, this cycle was named dry conditions. This was performed to ensure the measurements reproducibility and to eliminate the potential interference of water retained, in particular considering the hygroscopicity of the CIGS. For the measurements in wet conditions the samples were previously dissolved in bi-distilled water in the ratio 40:60 wt % (sample:water). The obtained paste was subsequently sandwiched between two gold electrodes. During the measurements in wet conditions, the sandwiched powder between the two electrodes were kept in a BDS 1308 liquid device, coupled to the spectrometer following the experimental process following a described procedure [47–50]. During the conductivity measurements, the temperature was maintained (isothermal experiments) with a stepwise of 20 °C from 20 to 200 °C controlled by a nitrogen jet (QUATRO from Novocontrol) with a temperature error of 0.1 K during every single sweep in frequency. From the frequency dependence of complex impedance $Z^*(\omega) = Z'(\omega) + j \cdot Z''(\omega)$, the real part of the conductivity is given as Equation (1)

$$\sigma'(\omega) = \frac{Z'(\omega) \cdot L}{\left[(Z'(\omega))^2 + (Z''(\omega))^2\right] \cdot S} = \frac{L}{R_0 \cdot S} \quad (1)$$

where L and S are the thickness and area of the sample sandwiched between the electrodes, respectively, and R_0 the resistance of the sample.

3. Results and Discussion

3.1. Synthesis of $CuIn_xCr_yGa_{1-x-y}Se_2$ Nanocrystals

CIGS nanocrystals were synthesized via the solvothermal method by mixing the corresponding amounts of the source materials (see Materials and methods for a full description of the synthesis), in this case Cu, Se, GaI$_3$, InCl$_3$, and Cr(ClO$_4$)$_3$, at the desired stoichiometry in ethylenediamine as solvent (Figure 1). In our methodology, the In/Ga ratio was varied from 0 to 1. The solution was then loaded into a Teflon autoclave container and annealed in an open-air oven for 36 h at 220 °C. The resultant powder was washed several with ethanol and water. Then, the solid centrifugated at 4000 rpm and ultra-sonicated to finally give a black nano-crystalline powder of $CuIn_xCr_yGa_{1-x-y}Se_2$.

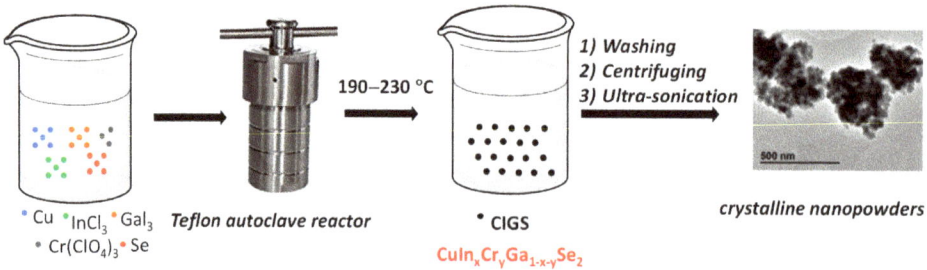

Figure 1. Schematic representation of the preparation of CIGS nanoparticles.

3.2. Structural Study

3.2.1. XRD of the Precursor Powder

Figure 2a shows the XRD pattern of the prepared sample with a chalcopyrite structure. Three sharp diffraction peaks were obtained, which confirms a chalcopyrite (tetragonal) phase structure corresponding to $CuIn_{0.4}Ga_{0.6}Se_2$. The diffraction peaks match well to standard CIGS file (PDF card no. 00–035–1101). The three peaks observed were identified as a pure phase of the $CuIn_xGa_{1-x}Se_2$ chalcopyrite phase with no secondary phases. The main peaks of CIGS powders correspond to crystallographic planes labelled as (112) at 27.36°, (204)/(220) at 2θ = 45.35°, and (116)/(312) at 2θ = 53.62°, respectively, in agreement with similar materials [51].

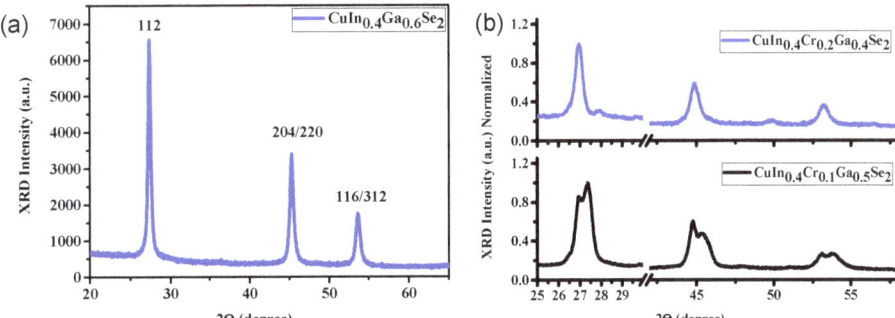

Figure 2. XRD patterns of (**a**) CuIn$_{0.4}$Ga$_{0.6}$Se$_2$ precursor nanocrystals and (**b**) CuIn$_{0.4}$Cr$_{0.1}$Ga$_{0.5}$Se$_2$ and CuIn$_{0.4}$Cr$_{0.2}$Ga$_{0.4}$Se$_2$ precursor nanocrystals both prepared by autoclave hydrothermal method.

Figure 2b displays the XRD pattern for CuIn$_{0.4}$Cr$_{0.1}$Ga$_{0.5}$Se$_2$ and CuIn$_{0.4}$Cr$_{0.2}$Ga$_{0.4}$Se$_2$ nano-powders. In the case of CuIn$_{0.4}$Cr$_{0.2}$Ga$_{0.4}$Se$_2$ powders, only the crystallographic peaks corresponding to chalcopyrite CIGS structure could be observed, i.e., (112) peak located at 26.94°, (204)/(220) peaks at 2θ = 44.88°, and (116)/(312) at 2θ = 53.22°, respectively. The main characteristic peaks for the CuIn$_{0.4}$Cr$_{0.1}$Ga$_{0.5}$Se$_2$ samples were located at 26.93°, 44.77°, and 53.03°, respectively, whereas they were combined with a subsidiary wurtzite Cu$_{2-x}$Se phase located at 27.35°, 45.48°, 53.79°, respectively, shifted to higher 2θ degrees [52]. On the other hand, the sample of CuIn$_{0.4}$Cr$_{0.1}$Ga$_{0.5}$Se$_2$ displayed the main characteristic peaks of chalcopyrite CIGS in addition to some wurtzite Cu$_{2-x}$Se peaks. Herein, every peak is built up by combination of two divided peaks like camel humps (like it has been doubleted), which can be attributed to the presence of a small percent of Cr(III) ions in the crystal lattice. The incorporation of ions with three electrons in the 3d shell and spin up, and the 3d^{10} configuration of Cu(I) ion results in one unique hole in the 3d shell, because the spin of the Cu(I) ion is aligned antiparallel to that of Cr(III) ion. These holes are supposed to be delocalized and occupy the states in a narrow d band. The metallic behavior is associated with the t_{2g} orbital of these delocalized Cu(I) holes. Therefore, we expect here a strong competition between Cu$_{2-x}$Se wurtzite phase formation and CuIn$_{0.4}$Cr$_{0.1}$Ga$_{0.5}$Se$_2$, which result in the formation of both phases.

In CuIn$_{0.4}$Cr$_{0.2}$Ga$_{0.4}$Se$_2$ samples, where the percentage of Cr(III) is equal to 20% which is higher than in CuIn$_{0.4}$Cr$_{0.1}$Ga$_{0.5}$Se$_2$ samples, the increase in Cr(III) content affects to the CuIn$_x$Cr$_y$Ga$_{1-x-y}$Se$_2$, (x = 0.4 and y = 0.0, 0.1, 0.2, 0.3) chalcopyrite phase formation. The CuIn$_{0.4}$Cr$_{0.3}$Ga$_{0.3}$Se$_2$ has 30% of Cr(III) ions in addition to Ga(III) ions, which leads to a competition between Cr(III) and Ga(III) ions rather than the stated competition between Cr(III) and Cu(I) ions.

The mean crystallite size of the prepared polycrystalline CIGS was calculated according to Scherer's formula (Equation (2))

$$D = \frac{K\lambda}{\beta \cos\theta} \quad (2)$$

where β is the full width at half maximum (FWHM), λ is the wavelength of X-rays (1.5418 Å), K is the Scherer's constant which depends on the crystallite shape and is close to 1, and θ is the Bragg angle at the center of the peak.

The obtained XRD patterns can be assigned to a tetragonal CuIn$_{0.4}$Ga$_{0.6}$Se$_2$ crystallographic phase (PDF card no. 00–035-1101) with a preferred orientation along the (112) plane.

No other stoichiometric composition of CIGS was obtained. The well-defined and sharp diffraction peaks indicated that the material showed good crystallinity and evidences the absence of any additional diffraction peaks of possible binary phases or impurities. Table 1 shows the measured position of the 112-diffraction peak, its FWHM, and the crystalline size for all CuInGaSe$_2$ with different contents of Cr. The crystalline size ranges from 10 to 20 nm depending on the Cr content in the powders.

Table 1. XRD parameters and crystallite size of CIGS nano-crystalline precursor powders

Sample Name	hkl	2θ (°)	FWHM	Crystallite Size (nm)
CuIn$_{0.4}$Ga$_{0.6}$Se$_2$	112	27.36	0.39	20.3 ± 0.6
CuIn$_{0.4}$Cr$_{0.1}$Ga$_{0.5}$Se$_2$	112	26.92	0.83	10.2 ± 1.2
CuIn$_{0.4}$Cr$_{0.2}$Ga$_{0.4}$Se$_2$	112	26.94	0.48	17.4 ± 0.8
CuIn$_{0.4}$Cr$_{0.3}$Ga$_{0.3}$Se$_2$	112	27.59	0.41	18.1 ± 0.5

3.2.2. XRD of CIGS Thin Films

After characterizing the CIGS nanoparticles, we evaluated their structure after deposition on indium tin oxide (ITO) glass substrates and fabrication of thin films by spin coating process. Figure 3a shows the XRD graph of the thin film CuIn$_{0.4}$Ga$_{0.6}$Se$_2$ on the ITO substrate. XRD patterns are consistent with chalcopyrite (tetragonal) crystal structure and exhibit a peak broadening due to their nanoscale crystal size. The diffraction peaks shift to lower 2θ degrees with decreasing Ga content by incorporation of Cr(III) ions in Ga(III) sites. The increase in the lattice spacing is due to smaller Cr atoms substituting for larger Ga atoms which in turn lead to a significant change in lattice parameters values.

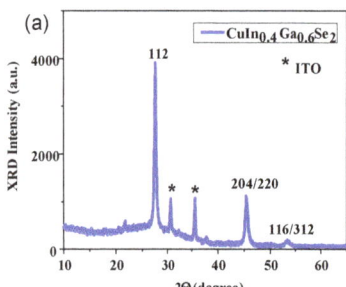
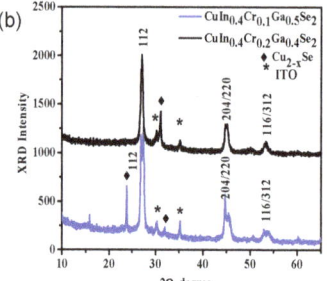
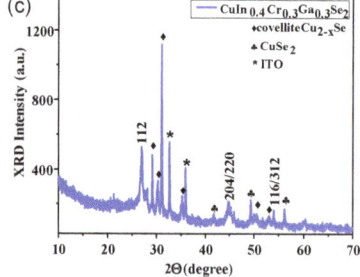

Figure 3. XRD patterns of thin film deposited onto ITO by spin coating process of (a) CuIn$_{0.4}$Ga$_{0.6}$Se$_2$ substrate, (b) CuIn$_{0.4}$Cr$_{0.1}$Ga$_{0.5}$Se$_2$ and CuIn$_{0.4}$Cr$_{0.2}$Ga$_{0.4}$Se$_2$, (c) CuIn$_{0.4}$Cr$_{0.3}$Ga$_{0.3}$Se$_2$.

The stoichiometric compound Cu(In,Ga)Se$_2$ crystallizes in the tetragonal chalcopyrite type crystal structure with space group *I-42d* (122). Within this crystal structure the monovalent cations of Cu(I) occupy the 4a site (0 0 0), the trivalent cations of In(III) and Ga(III) are located on the 4b position (0 0 $\frac{1}{2}$) and the selenium anions are on the 8d site (x $\frac{1}{4}\frac{1}{8}$). The cations are tetrahedrally coordinated by the anions and vice versa, substitution of Ga(III) ions with Cr affect lattice values due to the difference in atomic radius of both Cr and Ga atoms. In general, the absorbing layers of Cu(In,Ga)Se$_2$ exhibit a poor copper composition (Cu/(In + Ga) < 1), whereby the chalcopyrite-like crystalline structure persists together with the occupation of the sites of Wyckoff. This can cause specific changes due to possible defects in the material. As a consequence, there is a strong correlation between the concentration of these defects and the electronic and optical properties of the material, which can be especially interesting to adapt to high-efficiency photovoltaic devices. All the doped samples with Cr showed an enhancement in the lattice value corresponding to the FWHM value for CuIn$_{0.4}$Ga$_{0.4}$Se$_2$. We notice the main characteristic peaks of CuIn$_{0.4}$Ga$_{0.6}$Se$_2$ and those characteristics of ITO, which were located at 2θ = 30.58°

and 35.48°, respectively [53]. A well-defined crystallographic chalcopyrite structure with all preferred orientation a long 112, (204)/(220) and (116)/(312) for $CuIn_{0.4}Ga_{0.4}Se_2$ at $2\theta = 27.49°$, $2\theta = 45.37°$, and $2\theta = 53.36°$ is clearly observed.

The XRD patterns of the thin film for $CuIn_{0.4}Cr_{0.1}Ga_{0.5}Se_2$ and $CuIn_{0.4}Cr_{0.2}Ga_{0.4}Se_2$ on ITO substrate are displayed in Figure 3b. We notice those main characteristic peaks of $CuIn_{0.4}Cr_{0.2}Ga_{0.4}Se_2$ clearly at $2\theta = 27.18°$, $2\theta = 44.68°$, and $2\theta = 53.33°$ and those characteristics of ITO. Wurtzite $Cu_{2-x}Se$ phase was observed by thin film formation in both of $CuIn_{0.4}Cr_{0.1}Ga_{0.5}Se_2$ and $CuIn_{0.4}Cr_{0.2}Ga_{0.4}Se_2$, whereas it is observed in a small percent for $CuIn_{0.4}Cr_{0.2}Ga_{0.4}Se_2$ thin film samples. The $CuIn_{0.4}Cr_{0.1}Ga_{0.5}Se_2$ thin film structure characteristic peaks were observed at $2\theta = 26.72°$ and $27.36°$, $2\theta = 44.52°$ and $45.45°$, and $2\theta = 52.99°$ and $53.67°$ for 112, (204)/(220) and (116)/(312), respectively. The peaks duplicity is clearly observed for $CuIn_{0.4}Cr_{0.1}Ga_{0.5}Se_2$ thin film structure than those of $CuIn_{0.4}Cr_{0.2}Ga_{0.4}Se_2$. On the other hand the pure well defined highly oriented 112, (204)/(220) and (116)/(312) peaks for $CuIn_{0.4}Cr_{0.2}Ga_{0.4}Se_2$ were identified. As shown in Figure 3c, which shows the XRD graph of the thin film $CuIn_{0.4}Cr_{0.3}Ga_{0.3}Se_2$ on ITO substrate, the characteristic peaks of $CuIn_{0.4}Cr_{0.3}Ga_{0.3}Se_2$ at $2\theta = 26.96°$, $2\theta = 44.78°$, and $2\theta = 53.94°$ were observed. Mixture of copper selenide phases are very apparent in the XRD pattern for the $CuIn_{0.4}Cr_{0.3}Ga_{0.3}Se_2$ sample—i.e., covellite $Cu_{2-x}Se$ and $CuSe_2$—which is due to the stated confliction between both Cr(III) and Cu(I) ions at definite amount of incorporated Cr(III) ions that results in impurity phase formation structure. Using Scherrer formula the crystal size of all the prepared thin films of $CuIn_{0.4}Ga_{0.4}Se_2$, $CuIn_{0.4}Cr_{0.1}Ga_{0.5}Se_2$, $CuIn_{0.4}Cr_{0.2}Ga_{0.4}Se_2$, and $CuIn_{0.4}Cr_{0.3}Ga_{0.3}Se_2$ were calculated. It was found that the crystal size is equal to 20, 10, 17, and 18 nm for all the prepared samples of the structure where x = 0.4 and y = 0.0, 0.1, 0.2, 0.3, respectively. As inferred from Table 2, the FWHM of $CuIn_{0.4}Cr_{0.1}Ga_{0.5}Se_2$ thin film sample show remarkable higher value than those of $CuIn_{0.4}Ga_{0.6}Se_2$, $CuIn_{0.4}Cr_{0.2}Ga_{0.4}Se_2$, and $CuIn_{0.4}Cr_{0.3}Ga_{0.3}Se_2$. This remarkable increase of FWHM value is similar to those FWHM values of the grown samples, which confirms the structure defects and the stablished competition between Cr(III) and Cu(I) ions in the unit cell structure of those with low Cr% = 10, and Cr% = 30. The weak broad peak in the XRD pattern of this CIGS nanoparticle film is indexed to the (112) plane of a CIGS chalcopyrite crystal structure, which also implies that the film is very thin and not high in crystallinity due to its low annealing temperature [54].

Table 2. XRD parameters and crystallite size of CIGS nano-crystalline thin films for the 112 peak

Sample Name	2θ (°)	FWHM	Crystallite Size (nm)
$CuIn_{0.4}Ga_{0.6}Se_2$	27.5	0.40	20.2 ± 1.3
$CuIn_{0.4}Cr_{0.1}Ga_{0.5}Se_2$	27.0	0.80	10.2 ± 0.5
$CuIn_{0.4}Cr_{0.2}Ga_{0.4}Se_2$	27.2	0.50	17.1 ± 1.2
$CuIn_{0.4}Cr_{0.3}Ga_{0.3}Se_2$	26.7	0.44	18.0 ± 1.1

3.3. FE-SEM Analysis

3.3.1. FE-SEM Analysis of CIGS Nano-Crystalline Powders

The particle size and morphology of the resultant CIGS was evaluated by FE-SEM. Figure 4 shows FE-SEM images for $CuIn_xCr_yGa_{1-x-y}Se_2$ where x = 0.4 and y = (0.0, 0.1, 0.2, 0.3) nanocrystals (see Supplementary Figure S2 for images at 2 μm). According to FE-SEM images of the nano-crystalline precursor powders, $CuIn_{0.4}Ga_{0.6}Se_2$, $CuIn_{0.4}Cr_{0.1}Ga_{0.5}Se_2$, $CuIn_{0.4}Cr_{0.2}Ga_{0.4}Se_2$, and $CuIn_{0.4}Cr_{0.3}Ga_{0.3}Se_2$ contain agglomerates of spherical particles and some tetragonal and plate like shapes. Some tetrahedron hexagonal phases were determined in both $CuIn_{0.4}Cr_{0.1}Ga_{0.5}Se_2$, $CuIn_{0.4}Cr_{0.2}Ga_{0.4}Se_2$, and $CuIn_{0.4}Cr_{0.3}Ga_{0.3}Se_2$, in consistency with the XRD results. Those tetrahedron and hexagonal structures refer to the presence of $Cu_{2-x}Se$ and $CuSe_2$ phases in a complete matching with the XRD peaks referred to CuSe phases. Those tetrahedron and hexagonal structures refer to the presence of $Cu_{2-x}Se$ and $CuSe_2$ phases in a complete matching with the XRD peaks referred to CuSe phases [55]. For the FE-SEM images of $CuIn_{0.4}Ga_{0.6}Se_2$ nano-crystalline powders only those spherical of the chalcopyrite structure are present. No other morphologies have been observed, which in turn confirms the XRD data and of crystal structure and particle size calculations.

   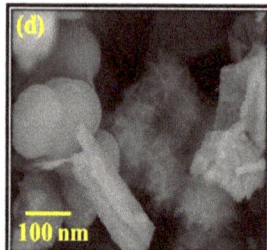

Figure 4. FE-SEM images for (**a**) $CuIn_{0.4}Ga_{0.6}Se_2$, (**b**) $CuIn_{0.4}Cr_{0.1}Ga_{0.5}Se_2$, (**c**) $CuIn_{0.4}Cr_{0.2}Ga_{0.4}Se_2$, and (**d**) $CuIn_{0.4}Cr_{0.3}Ga_{0.3}Se_2$ nano-crystalline powders at 100 nm.

The structures of the prepared precursor powders provide a promising feedback about the expected thin films surface morphology, which resemble the nano crystalline precursor powders. The average particle size is approximately (10–20 nm) in accordance with the HRTEM and XRD crystallite size. Such characteristic properties will play a vital role in the deposition process of the CIGS thin films. Particle size, morphology, uniformity, homogeneity, adhesion and cohesion of the prepared precursor powders; all these properties will affect the nature of the resultant films by any formation process. In general, one aims for particulate precursor powders with a few hundred nanometers and good uniformity in particles shape will lead to a sufficient adhesion to the substrate and sufficient cohesion of one particle to another, as shown in Figure 4.

3.3.2. FE-SEM Analysis of CIGS Thin Films

FE-SEM images for the prepared thin films containing $CuIn_{0.4}Ga_{0.6}Se_2$, $CuIn_{0.4}Cr_{0.1}Ga_{0.5}Se_2$, $CuIn_{0.4}Cr_{0.2}Ga_{0.4}Se_2$, and $CuIn_{0.4}Cr_{0.3}Ga_{0.3}Se_2$ at 20 μm are shown in Figure 5. All films displayed a homogenous nano-crystalline particle distribution after deposition using the spin-coating process.

Figure 5. FE-SEM images for (**a**) $CuIn_{0.4}Ga_{0.6}Se_2$, (**b**) $CuIn_{0.4}Cr_{0.1}Ga_{0.5}Se_2$, (**c**) $CuIn_{0.4}Cr_{0.2}Ga_{0.4}Se_2$, and (**d**) $CuIn_{0.4}Cr_{0.3}Ga_{0.3}Se_2$ thin films at 20 μm.

3.4. HR-TEM Analysis

Figure 6 shows HR-TEM images of the $CuIn_{0.4}Ga_{0.6}Se_2$ at different magnification scales. The d-spacing between crystallographic planes is shown in Figure 6g,h, and corresponds to 3.32 Å. This lattice spacing corresponds to tetragonal $CuIn_xGa_{1-x}Se_2$ in accordance with the XRD peaks of the chalcopyrite $CuIn_xGa_{1-x}Se_2$ [56].

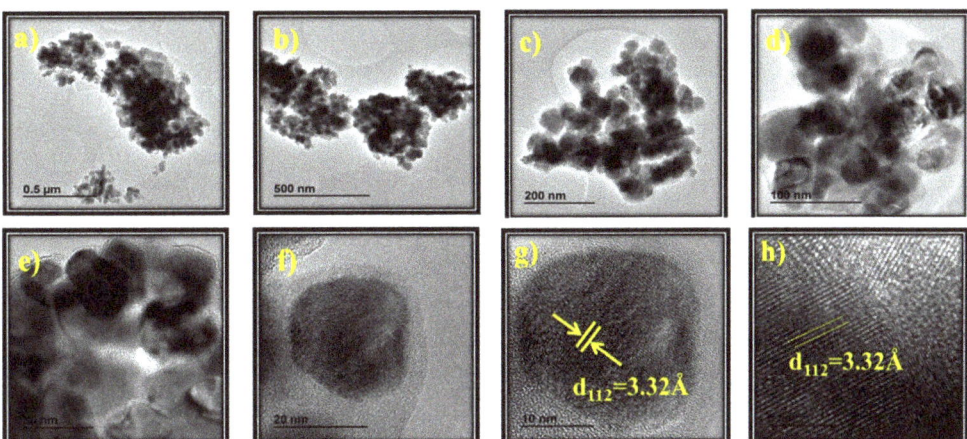

Figure 6. (a–f) HR-TEM images for CuIn$_{0.4}$Ga$_{0.6}$Se$_2$ nanocrystals and (g,h) d-spacings of chalcopyrite (tetragonal) CuIn$_{0.4}$Ga$_{0.6}$Se$_2$.

Therefore, we can conclude that the crystallinity of the tetragonal chalcopyrite CuIn$_{0.4}$Ga$_{0.6}$Se$_2$ maintains a structure basically composed by nanocrystals of approximately less than 20 nm in diameter. The d-spacings observed in HR-TEM images are also consistent with tetragonal CuIn$_x$Ga$_{1-x}$Se$_2$ chalcopyrite phase and no other crystal phases were observed in accordance with the XRD pattern of the prepared powder. The HR-TEM images show well defined small grains of a few tens of nanometers in agreement with the observed results by FE-SEM (Figure 5). The measured lattice spacing of 3.32 Å, observed in the highest magnification HR-TEM images (Figure 6g,h), matches well with the interplanar distance between the (112) crystallographic planes.

3.5. EDX Analysis

Energy dispersive X-ray spectroscopy (EDX) analysis of the CIGS nanocrystals was performed, and the average atomic ratios are shown in Table 3. EDX results reveal that the prepared nano-crystalline powders contain certain amount of the constituent elements in accordance with the desired structures (Supplementary Figure S3). The measured stoichiometric atomic ratios 0.87:0.4:0.6:1.98 for copper, indium, gallium, and selenium obtained by EDX are closer to the expected. However, a slight decrease of Cu and Se was found in the resultant composition in addition to an observed decrease in Cu content. The slightly copper-poor composition is beneficial for the formation of *p*-type semiconductor. The proposed composition of the prepared nanocrystal precursor powder in the first sample has a molar Cu/In/Ga/Se ratio of 1.0:0.4:0.6:2.0 with a fixed In ratio for all prepared series with incorporation of Cr atoms in Ga sites.

Table 3. Composition of nano-crystalline precursor powders according to EDX analysis

Sample	Cu (%)	In (%)	Cr (%)	Ga (%)	Se (%)
CuIn$_{0.4}$Ga$_{0.6}$Se$_2$	22.6	9.7	-	16.3	51.4
CuIn$_{0.4}$Cr$_{0.1}$Ga$_{0.5}$Se$_2$	29.4	8.8	1.6	11.5	48.7
CuIn$_{0.4}$Cr$_{0.2}$Ga$_{0.4}$Se$_2$	26.9	9.9	3.2	9.9	50.1
CuIn$_{0.4}$Cr$_{0.3}$Ga$_{0.3}$Se$_2$	33.6	7.2	4.9	6.1	48.2

We observed a copper deficiency from 1.0 to 0.87, which indicates that the prepared CuIn$_{0.4}$Ga$_{0.6}$Se$_2$ are *p*-type materials. A comparison between the proposed precursor composition and the measured EDX composition is indicated in Table 3. Although the control of the composition ratios in this type of chalcopyrites is difficult, we have been able, with this hydrothermal autoclave method, to optimize the resultant product ratios of In and Ga by controlling several factors such as metal source, time of stirring and preparation temperature. For $CuIn_{0.4}Cr_{0.1}Ga_{0.5}Se_2$ sample the proposed composition of the prepared precursor (atomic ratio of Cu/In/Cr/Ga/Se) is equal to 1:0.4:0.1:0.5:2 with an atomic percentage equal 25:10:2.5:12.5:50. The resultant atomic ratio was equal to 1.3:0.4:0.07:0.53:2.3 with an atomic ratio equal 29.44:8.8:1.58:11.49:48.69 for Cu, In, Cr, Ga, and Se, respectively. A slight increase in both Cu and Se atomic ratios is observed which in accordance with the XRD results for the presence of wurtzite $Cu_{2-x}Se$ phase in competition with the $CuIn_{0.4}Cr_{0.1}Ga_{0.5}Se_2$ chalcopyrite phase formation. This can be supposed also to the formation of $CuIn_{0.4}Cr_{0.1}Ga_{0.5}Se_2$ quartzite phase also noticed, but in total Ga + Cr = Ga main atomic ratio before substitution and In ratio is completely fixed which in turn does not affect the chalcopyrite phase formation even in the formation of the subsidiary phase as we explained in XRD part.

For $CuIn_{0.4}Cr_{0.2}Ga_{0.4}Se_2$ we noticed that the resulted composition atomic ratios for Cu, In, Cr, Ga, and Se were equal to 26.9:9.9:3.2:9.9:50.1 with an atomic ratio equal to (1.2:0.43:0.14:0.43:2.3). The proposed composition for this structure were (1:0.4:0.2:0.4:2) with an atomic ratio equal to (25:10:5:10:50). It is noticed that an equality in both In and Ga atomic ratios, slight decrease in Cr content than proposed it should work as 20% atomic ratio and it was found equal to 14%. This decrease in Cr content does not affects the chalcopyrite crystal structure, in contrast, it gives a stability with preferred orientation to the structure.

Mixture of different copper selenide phases appear as secondary phases with the determined chalcopyrite tetragonal structure of $CuIn_{0.4}Cr_{0.3}Ga_{0.3}Se_2$. The proposed composition for this structure were (1:0.4:0.3:0.3:2) with an atomic ratio equal to (25:10:7.5:7.5:50). Herein, equal amounts of Ga and Cr atoms were incorporated together in the studied phase with a noticeable variation in both Cu and Se atomic ratios, respectively. We found that the composition for Cu, In, Cr, Ga, and Se was equal to 33.60:7.22:4.90:6.09:48.19 with an atomic ratio of (1.8:0.4:0.27:0.33:2.6).

3.6. Optical Properties

Optical properties have been investigated from the transmittance spectra. All the $CuIn_xCr_yGa_{1-x-y}Se_2$ thin films exhibited broad absorption in the visible region. The absorption coefficients were obtained for thin films samples prepared using spin-coating process with nanocrystals.

Near the absorption edge or in the strong absorption zone of the transmittance spectra of materials, the absorption coefficient is related to the optical energy gap, E_g, which can be determined by the Tauc's equation [57], Equation (3)

$$\alpha = \frac{A(h\nu - E_g)^n}{h\nu} \quad (3)$$

where A is a constant, h is the Planck constant, ν is frequency, and n is an index that characterizes the optical absorption process and is equal to 2 for direct allowed transitions and 0.5 for indirect allowed transitions.

The Tauc's plot for $CuIn_xCr_yGa_{1-x-y}Se_2$ thin films are displayed in Figure 7. The shape of Tauc's plot indicates that the deposited $CuIn_xCr_yGa_{1-x-y}Se_2$ thin films possessed a direct band gap. Extrapolation of the straight line to zero absorption coefficient ($\alpha = 0$) allows an estimation of E_g. For that, the band gaps were obtained by plotting $(\alpha h\nu)^2$ vs. the energy in eV and extrapolating the linear part of the spectrum ($h\nu$).

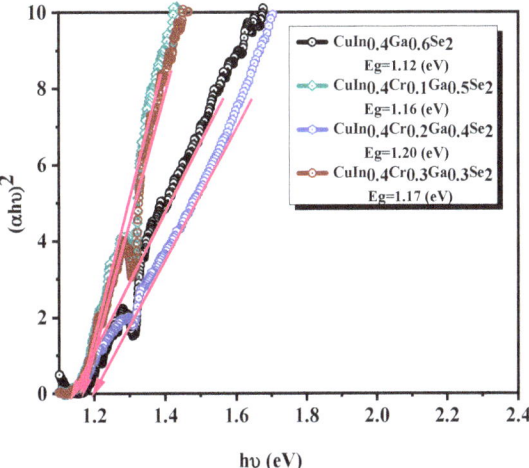

Figure 7. Tauc's plot for $CuIn_{0.4}Ga_{0.6}Se_2$, $CuIn_{0.4}Cr_{0.1}Ga_{0.5}Se_2$, $CuIn_{0.4}Cr_{0.2}Ga_{0.4}Se_2$, and $CuIn_{0.4}Cr_{0.3}Ga_{0.3}Se_2$ precursor powders.

The prepared $CuIn_{0.4}Ga_{0.6}Se_2$, $CuIn_{0.4}Cr_{0.1}Ga_{0.5}Se_2$, $CuIn_{0.4}Cr_{0.2}Ga_{0.4}Se_2$, and $CuIn_{0.4}Cr_{0.3}Ga_{0.3}Se_2$ thin films with E_g of 1.12, 1.16, 1.20, and 1.17 eV, respectively, which are close to the optimum value for solar photoelectric conversion of 1.5 eV. The differences of the band gaps and absorption spectra of the thin films was caused by the changing particle size and morphology of the prepared $CuIn_{0.4}Ga_{0.6}Se_2$, $CuIn_{0.4}Cr_{0.1}Ga_{0.5}Se_2$, $CuIn_{0.4}Cr_{0.2}Ga_{0.4}Se_2$, and $CuIn_{0.4}Cr_{0.3}Ga_{0.3}Se_2$ thin films with composition difference. Although composition dependency of E_g has been observed for other semiconductor particles like CZTS and $CuInS_2$, there has been scarce investigation of the influence of the particle composition on light−electricity conversion efficiency, especially for those new prepared $CuIn_{0.4}Ga_{0.6}Se_2$, $CuIn_{0.4}Cr_{0.1}Ga_{0.5}Se_2$, $CuIn_{0.4}Cr_{0.2}Ga_{0.4}Se_2$, and $CuIn_{0.4}Cr_{0.3}Ga_{0.3}Se_2$ thin films.

3.7. Dielectric Spectra Analysis

In order to get information about the behavior of the ionic conductivity, electrochemical impedance spectroscopy measurements were performed on the powdered solid compounds, namely, $CuIn_{0.4}Cr_{0.1}Ga_{0.5}Se_2$, $CuIn_{0.4}Cr_{0.2}Ga_{0.4}Se_2$, and $CuIn_{0.4}Cr_{0.3}Ga_{0.3}Se_2$. The measurements were carried out at different temperatures (from 20 to 200 °C) to obtain information on the samples' conductivity. The study of the conductivity was analyzed in terms of the corresponding Bode diagrams, which are shown in Figure 8 (dry conditions) and Figure 9 (wet conditions), respectively. These plots display the variation of the conductivity with the frequency, by. plotting the double logarithmic plot of the conductivity in S/cm versus frequency in (Hz) of each sample in all range of temperatures.

In the measurements in dry conditions (Figure 8), the conductivity σ' is characterized in the Bode plot by a plateau, where the phase angle tends to zero, the imaginary part of the impedance will be zero, and then the corresponding conductivity represents the direct-current conductivity (σdc) of the sample. A close inspection at Figure 8 revealed that samples $CuIn_{0.4}Cr_{0.1}Ga_{0.5}Se_2$, $CuIn_{0.4}Cr_{0.2}Ga_{0.4}Se_2$, and $CuIn_{0.4}Cr_{0.3}Ga_{0.3}Se_2$ possess a conductivity, practically constant in all the range of frequencies for all temperatures under study, which is a typical behavior for a conductor. Similar results have been observed in samples of multilayer graphene in polypropylene nanocomposites [58]. This can be explained as a Debye relaxation due to the motion and reorientation of the dipoles and localized charges as consequence of the electric field applied, which dominates the dc-conductivity [59,60]. For all temperatures under study, conductivity increases with the temperature following the trend, σ′ (CuIn$_{0.4}$Ga$_{0.6}$Se$_2$) > σ′ (CuIn$_{0.4}$Cr$_{0.2}$Ga$_{0.4}$Se$_2$) > σ′ (CuIn$_{0.4}$Cr$_{0.1}$Ga$_{0.5}$Se$_2$) > σ′ (CuIn$_{0.4}$Cr$_{0.3}$Ga$_{0.3}$Se$_2$).

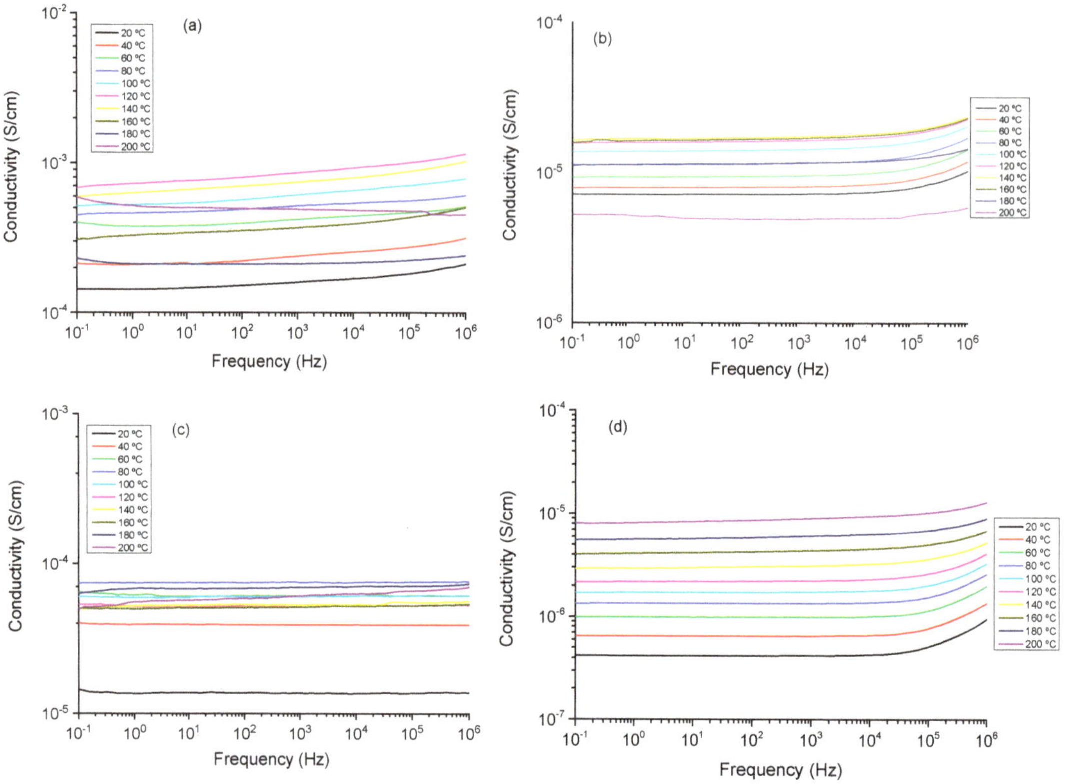

Figure 8. Double logarithmic plot of the real part of the conductivity versus frequency for (**a**) CuIn$_{0.4}$Ga$_{0.6}$Se$_2$, (**b**) CuIn$_{0.4}$Cr$_{0.1}$Ga$_{0.5}$Se$_2$, (**c**) CuIn$_{0.4}$Cr$_{0.2}$Ga$_{0.4}$Se$_2$, and (**d**) CuIn$_{0.4}$Cr$_{0.3}$Ga$_{0.3}$Se$_2$ precursor powders obtained in dry conditions.

As shown in measurements in wet conditions (Figure 9), a deviation from σdc in the spectrum of the conductivity can be seen at low frequency values. This effect is attributed to the electrode polarization (EP) effect because of the blocking electrodes, and produced by mobile charge accumulation. However, it is only observed in the measurements realized below wet conditions [61,62]. The conductivity also exhibits a phenomenon of dispersion that obeys a behavior described by Jonscher law [63,64], given by: σ(ω) = σdc +σac, being σdc the dc conductivity and the alternating-current σdc = Aω^m, where A is a factor dependent of the temperature and m for an ideal Debye dielectric and ideal ionic type crystals are 1 and 0, respectively.

Experimental data of the conductivity were fitted to the Jonscher law and the parameters are shown in Table 4. As we can see in Figure 10, only two different regions are present. On the one hand, at moderate and low frequencies where the dc-conductivity will be determined (Cond), and the region of sub-diffusive conductivity (SD), observed in the interval between 10^5 and 10^7 Hz for the samples CuIn$_{0.4}$Ga$_{0.6}$Se$_2$, CuCr$_{0.1}$In$_{0.4}$Ga$_{0.5}$Se$_2$, and CuCr$_{0.3}$In$_{0.4}$Ga$_{0.3}$Se$_2$ respectively. Is important highlight that the sample, CuCr$_{0.2}$In$_{0.4}$Ga$_{0.4}$Se$_2$, displays a behavior of a pure conductor in all the range of frequencies under all temperatures under study [65–67].

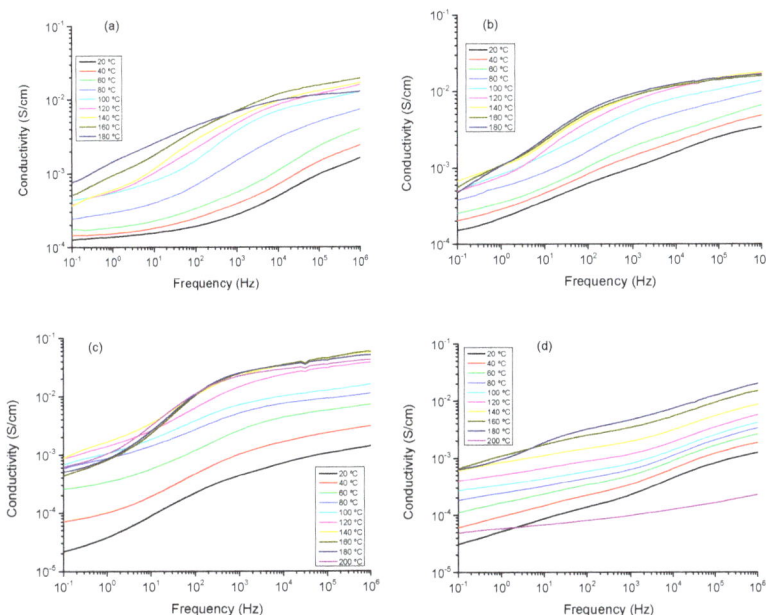

Figure 9. Double logarithmic plot of the real part of the conductivity versus frequency for (**a**) CuIn$_{0.4}$Ga$_{0.6}$Se$_2$, (**b**) CuIn$_{0.4}$Cr$_{0.1}$Ga$_{0.5}$Se$_2$, (**c**) CuIn$_{0.4}$Cr$_{0.2}$Ga$_{0.4}$Se$_2$, and (**d**) CuIn$_{0.4}$Cr$_{0.3}$Ga$_{0.3}$Se$_2$ precursor powders obtained in wet conditions.

Table 4. Jonscher parameters obtained by fitting the equation $\sigma(\omega) = \sigma_0 + A\omega^m$ to conductivity data, for some indicated temperature in all studied samples

Cr = 0.0	σ_{dc} [S cm^{-1}]	A	m	Cr = 0.1	σ_{dc} [S cm^{-1}]	A	m
T = 20 °C	1.6×10^{-4}	$10^{-6.96}$	0.39	T = 20 °C	7.4×10^{-6}	$10^{-10.5}$	0.73
T = 60 °C	4.2×10^{-4}	$10^{-6.20}$	0.32	T = 60 °C	9.8×10^{-6}	$10^{-9.98}$	0.68
T = 100 °C	6.1×10^{-4}	$10^{-5.68}$	0.29	T = 100 °C	1.5×10^{-5}	$10^{-9.07}$	0.57
T = 160 °C	3.7×10^{-4}	$10^{-6.17}$	0.34	T = 160 °C	1.8×10^{-5}	$10^{-9.70}$	0.66
Cr = 0.2	σ_{dc} [S cm^{-1}]	A	m	Cr = 0.3	σ_{dc} [S cm^{-1}]	A	m
T = 20 °C	1.4×10^{-5}	0	0	T = 20 °C	4.3×10^{-7}	$10^{-11.59}$	0.77
T = 60 °C	6.2×10^{-5}	0	0	T = 60 °C	1.0×10^{-6}	$10^{-10.97}$	0.72
T = 100 °C	6.1×10^{-5}	0	0	T = 100 °C	1.8×10^{-6}	$10^{-10.58}$	0.69
T = 160 °C	5.3×10^{-5}	0	0	T = 160 °C	4.5×10^{-6}	$10^{-9.89}$	0.62

In our samples, the values fitted for the parameter m, takes values between 0.4 and 1 for all the samples except in case of sample CuCr$_{0.2}$In$_{0.4}$Ga$_{0.4}$Se$_2$, where the value was 0, corresponding to an ideal conductor type in all the range of temperatures. For the other samples, the values obtained from the fit in the high frequency region which can be due to the reorientation motion of dipoles and more likely to the motion of the localized charges, which in the beginning are dominates over the dc-conductivity [65–67]. From Figure 9, we can see that the conductivity is a function of the amount of fillers that we have incorporated in the powdered. On the other hand, for CuCr$_{0.2}$In$_{0.4}$Ga$_{0.4}$Se$_2$, nanocomposite we observe that conductivity is practically constant in all the range of frequencies, only at frequencies higher than 10^6 Hz and some temperatures, the behavior of the sample shown a cut-off frequency where it starts increasing with the frequency. For the other samples, we can see that the real part of the conductivity is constant at the low frequencies region until a cut-off frequency where it starts increasing with the frequency, as if the sample were a capacitor. The value of σ constant means that the CIGS has only a resistive contribution and its value represents the electrical conductivity of the sample. In Table 4, the values of the conductivity for the samples and the Jonscher parameters are shown.

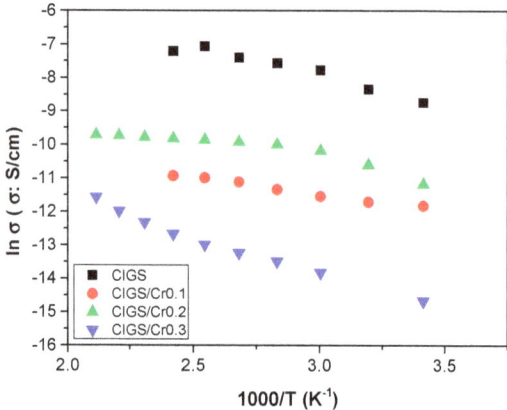

Figure 10. Temperature dependence of conductivity obtained from Bode diagram for all the samples studied.

As displayed in Figure 9, which shows the conductivity in wet conditions, it can be observed the temperature variations of the conductivity. In this plot, we can see at low temperatures σ_{dc} depends notably on the frequency and this effect tends to disappear when the temperature increases, but in our samples this behavior is not well developed due to the non-reproducibility of the measurements in wet conditions. As expected, the conductivity of all samples is strongly humidity dependent [65], increasing around two orders of magnitude, depending on the sample at each temperature. For example, for the sample $CuCr_{0.2}In_{0.4}Ga_{0.4}Se_2$ the conductivity at 60 °C is around 6.2×10^{-5} S/cm, however in wet conditions at the same temperature the conductivity was 5.7×10^{-3} S/cm. In all the range of temperatures studied the conductivity increase with the temperature following the trend: σ' ($CuIn_{0.4}Ga_{0.6}Se_2$) > σ' ($CuIn_{0.4}Cr_{0.2}Ga_{0.4}Se_2$) > σ' ($CuIn_{0.4}Cr_{0.1}Ga_{0.5}Se_2$) > σ' ($CuIn_{0.4}Cr_{0.3}Ga_{0.3}Se_2$).gure 10 shows the dependence of the conductivity measured in wet conditions with temperature. From this figure, we can see that conductivity follows a typical Arrhenius behavior with two different behaviors: one where the conductivity increases with increasing temperature, between 20 and 120 °C, following the Equation (4)

$$\ln \sigma_{dc} = \ln \sigma_\infty - \frac{E_{act}}{RT} \quad (4)$$

and other behavior for temperatures above 120 °C, where the conductivity begins to decrease due possibly to the dehydration of the sample.

From the slopes of the fits obtained following Equation (4), we calculated the activation energy (E_{act}) associated to the conduction process. The values obtained followed the trend $E_{act}(CuCr_{0.2}In_{0.4}Ga_{0.4}Se_2) = (5.5 \pm 0.2)$ kJ/mol < $E_{act}(CuCr_{0.1}In_{0.4}Ga_{0.4}Se_2) = (7.60 \pm 0.15)$ kJ/mol < $E_{act}(CuCr_{0.3}In_{0.4}Ga_{0.4}Se_2) = (18.8 \pm 0.2)$ kJ/mol < $E_{act}(CuIn_{0.4}Ga_{0.4}Se_2) = (24.8 \pm 0.4)$ kJ/mol. These results indicate that all the CIGS doped by chromium have low activation energy than $CuIn_{0.4}Ga_{0.4}Se_2$ thin films being the best doping percentage of Cr equal to 20% the optimum.

4. Conclusions

In conclusion, fabrication of low-cost chalcopyrite CIGS semiconductor material using Cr as a dopant yielded an excellent well-arranged crystalline structure with the same properties and small quantities of the high-cost Ga content. Incorporation of the Cr metal, as a doping agent, in the preparation solution of the metallic salt mixture stands as a pre-deposition process type of incorporation. Resulted precursor powders in nano-sized range need to be treated in a good manner which ensures formation of excellent adhesive solution. Spin coating process was applied to get the desired new doped Cr:CIGS chalcopyrite thin films. Electrochemical impedance spectroscopy measurements confirmed that those prepared doped Cr:CIGS with different percentages are promising semiconductor materials, as inferred from the experimental data which showed that the conductivity values increase with the temperature following the trend σ' ($CuIn_{0.4}Ga_{0.6}Se_2$) > σ' ($CuIn_{0.4}Cr_{0.2}Ga_{0.4}Se_2$) > σ' ($CuIn_{0.4}Cr_{0.1}Ga_{0.5}Se_2$) > σ' ($CuIn_{0.4}Cr_{0.3}Ga_{0.3}Se_2$). The activation energy of the powder $CuCr_{0.2}In_{0.4}Ga_{0.4}Se_2$ was 5.5 ± 0.2 kJ/mol and consequently more stable in all the range of temperatures; therefore, the optimal doping percentage of Cr was determined to be equal at 20%. Therefore, $CuIn_{0.4}Cr_{0.2}Ga_{0.4}Se_2$ thin film can be an excellent material for solar cell applications.

Supplementary Materials: The following are available online at https://www.mdpi.com/article/10.3390/nano11051093/s1, Supplementary Figure S1: Spin-coating process (feeding of dissolved metal solution onto ITO substrate using micropipette); Supplementary Figure S2: FE-SEM images for (a) $CuIn_{0.4}Ga_{0.6}Se_2$, (b) $CuIn_{0.4}Cr_{0.1}Ga_{0.5}Se_2$, (c) $CuIn_{0.4}Cr_{0.2}Ga_{0.4}Se_2$, and (d) $CuIn_{0.4}Cr_{0.3}Ga_{0.3}Se^2$ nanocrystalline precursor powders at 2 μm; Supplementary Figure S3: EDX charts for (a) $CuIn_{0.4}Ga_{0.6}Se_2$, (b) $CuIn_{0.4}Cr_{0.1}Ga_{0.5}Se_2$, (c) $CuIn_{0.4}Cr_{0.2}Ga_{0.4}Se_2$, and (d) $CuIn_{0.4}Cr_{0.3}Ga_{0.3}Se_2$ precursor powders.

Author Contributions: Conceptualization, B.M. and V.C.; Methodology, S.S., A.A., J.E. and M.A.A.; Validation, N.K., A.E. and A.E.N.; Formal analysis, M.A.A.; Investigation, S.S., A.A., J.E.; Resources, B.M. and V.C.; Data curation, A.A. and V.C.; Writing—original draft preparation, S.S. and M.A.A.; Writing—review and editing, J.E., B.M. and V.C.; Supervision, B.M. and V.C.; Project administration, B.M. and V.C.; Funding acquisition, B.M. and V.C. All authors have read and agreed to the published version of the manuscript.

Funding: This research has been supported by the Culture Affairs and Missions Sector, Ministry of Higher Education and Scientific Research (Egypt) and Ministerio de Economia y Competitividad (ENE2016-77798-C4-2-R), (ENE2015-69203-R) and Generalitat Valenciana (Prometeus 2014/044).

Data Availability Statement: The data presented in this study are available on request from the corresponding author.

Conflicts of Interest: The authors declare no conflict of interest.

References

1. Shubbak, M.H. Advances in solar photovoltaics: Technology review and patent trends. *Renew. Sustain. Energy Rev.* **2019**, *115*, 109383. [CrossRef]
2. Green, M.A.; Emery, K.; Hishikawa, Y.; Warta, W.; Dunlop, E.D.; Levi, D.H.; Ho-Baillie, A.W.Y. Solar cell efficiency tables (version 49). *Prog. Photovolt.* **2017**, *25*, 3–13. [CrossRef]
3. Gloeckler, M.; Sankin, I.; Zhao, Z. CdTe solar cells at the threshold to 20% efficiency. *IEEE J. Photovol.* **2013**, *3*, 1389–1393. [CrossRef]
4. Ramanujam, J.; Singh, U.P. Copper indium gallium selenide based solar cells—A review. *Energy Environ. Sci.* **2017**, *10*, 1306–1319. [CrossRef]
5. Jung, H.S.; Han, G.S.; Park, N.-G.; Ko, M.J. Flexible Perovskite Solar Cells. *Joule* **2019**, *3*, 1850–1880. [CrossRef]
6. Morales-Acevedo, A. Can we improve the record efficiency of CdS/CdTe solar cells? *Sol. Energy Mater. Sol. Cells* **2006**, *90*, 2213–2220. [CrossRef]
7. Ibn-Mohammed, T.; Koh, S.C.L.; Reaney, I.M.; Acquaye, A.; Schileo, G.; Mustapha, K.B.; Greenough, R. Perovskite solar cells: An integrated hybrid lifecycle assessment and review in comparison with other photovoltaic technologies. *Renew. Sustain. Energy Rev.* **2017**, *80*, 1321–1344. [CrossRef]
8. Meyer, B.; Klar, P. Sustainability and renewable energies–a critical look at photovoltaics. *Phys. Status Solidi RRL* **2011**, *5*, 318–323. [CrossRef]
9. Jackson, P.; Wuerz, R.; Hariskos, D.; Lotter, E.; Witte, W.; Powalla, M. Efects of heavy alkali elements in Cu (In,Ga)Se$_2$ solar cells with efficiencies up to 22.6%. *Phys. Status Solidi RRL* **2016**, *10*, 583–586. [CrossRef]
10. Hossain, M.; Alharbi, F. Recent advances in alternative material photovoltaics. *Mater. Technol.* **2013**, *28*, 88–97. [CrossRef]
11. Green, M.A. The Path to 25% Silicon Solar Cell Efficiency: History of Silicon Cell Evolution. *Prog. Photovol.* **2009**, *17*, 183–189. [CrossRef]

12. McFarland, E. Solar energy: Setting the economic bar from the top-down. *Energy Environ. Sci.* **2014**, *7*, 846–854. [CrossRef]
13. Andersson, B.A. Materials availability for large-scale thin-flm photovoltaics. *Prog. Photovolt. Res. Appl.* **2000**, *8*, 61–76. [CrossRef]
14. Ghosh, A.; Krishnan, Y. At a long-awaited turning point. *Nat. Nanotechnol.* **2014**, *9*, 491–494. [CrossRef] [PubMed]
15. Alharbia, F.; Bass, J.D.; Salhi, A.; Alyamani, A.; Kim, H.-C.; Miller, R.D. Abundant non-toxic materials for thin film solar cells: Alternative to conventional materials. *Renew. Energy* **2011**, *36*, 2753–2758. [CrossRef]
16. Rafique, S.; Abdullah, S.M.; Sulaiman, K.; Iwamoto, M. Fundamentals of bulk heterojunction organic solar cells: An overview of stability/degradation issues and strategies for improvement. *Renew. Sustain. Energy Rev.* **2018**, *84*, 43–53. [CrossRef]
17. Hegedus, S.S.; Shafarman, W.N. Thin-Film Solar Cells: Device, Measurements and Analysis. *Prog. Photovolt. Res. Appl.* **2004**, *12*, 155–176. [CrossRef]
18. Green, M.A. Silicon photovoltaic modules: A brief history of the first 50 years. *Prog. Photovolt. Res. Appl.* **2005**, *13*, 447–455. [CrossRef]
19. Goetzberger, A.; Hebling, C.; Schock, H.-W. Photovoltaic materials, history, status and outlook. *Mater. Sci. Eng. R Rep.* **2003**, *40*, 1–46. [CrossRef]
20. Repins, I.; Contreras, M.A.; Egaas, B.; DeHart, C.; Scharf, J.; Perkins, C.L.; To, B.; Noufi, R. 19.9%-efficient ZnO/CdS/CuInGaSe$_2$ Solar Cell with 81.2% Fill Factor. *Prog. Photovolt. Res. Appl.* **2008**, *16*, 235–239. [CrossRef]
21. Green, M.A.; Emery, K.; Hishikawa, Y.; Warta, W.; Dunlop, E.D. Solar cell efficiency tables (version 42). *Prog. Photovolt. Res. Appl.* **2013**, *21*, 827–837. [CrossRef]
22. Barbosa, P.; Rosero-Navarro, N.C.; Fa-Nian, S.; Figuereido, F.M.L. Protonic Conductivity of Nanocrystalline Zeolitic Imidazolate Framework 8. *Electrochim. Acta* **2015**, *153*, 19–27. [CrossRef]
23. Mousavi, S.H.; Müller, T.S.; Karos, R.; de Oliveira, P.W. Faster synthesis of CIGS nanoparticles using a modified solvothermal method. *J. Alloys Compd.* **2016**, *659*, 178–183. [CrossRef]
24. Wang, C.; Shei, S.; Chang, S. Novel solution process for synthesis of CIGS nanoparticles using polyetheramine as solvent. *Mater. Lett.* **2014**, *122*, 52–54. [CrossRef]
25. Mousavi, S.H.; Müller, T.S.; Oliveira, P. Synthesis of colloidal nanoscaled copper–indium–gallium–selenide (CIGS) particles for photovoltaic applications. *J. Colloid Interface Sci.* **2012**, *382*, 48–52. [CrossRef]
26. Gu, S.-I.; Shin, H.-S.; Yeo, D.-H.; Hong, Y.-W.; Nahm, S. Synthesis of the single phase CIGS particle by solvothermal method for solar cell application. *Curr. Appl. Phys.* **2011**, *11*, S99–S102. [CrossRef]
27. Mahboob, S.; Malik, S.; Haider, N.; Nguyen, C.; Malik, M.; O'Brien, P. Deposition of binary, ternary and quaternary metal selenide thin films from di isopropyl diseleno phosphinato-metal precursors. *J. Cryst. Growth* **2014**, *394*, 39–48. [CrossRef]
28. Chun, Y.G.; Kim, K.H.; Yoon, K.-H. Synthesis of CuInGaSe2 nanoparticles by solvothermal route. *Thin Solid Films* **2005**, *480–481*, 46–49. [CrossRef]
29. Bhattacharya, N.; Oh, M.-K.; Kim, Y. CIGS-based solar cells prepared from electro deposited precursor films. *Sol. Energy Mater. Sol. Cells* **2012**, *98*, 198–202. [CrossRef]
30. Woo, J.; Yoon, H.; Cha, J.; Jung, D.Y.; Yoon, S.S. Electrostatic spray-deposited CuInGaSe$_2$ nanoparticles: Effects of precursors'Ohnesorge number substrate temperature, and flow rate on thinfilm characteristics. *J. Aerosol Sci.* **2012**, *54*, 1–12. [CrossRef]
31. Kaelin, M.; Rudmanna, D.; Kurdesaua, F.; Zogg, H.; Meyer, T.; Tiwari, A.N. Low-cost CIGS solar cells by paste coating and selenization. *Thin Solid Films* **2005**, *480–481*, 486–490. [CrossRef]
32. Lee, E.; Joo, O.; Yoon, S.; Min, B.K. Synthesis of CIGS absorber layers via a paste coating. *J. Cryst. Growth* **2009**, *311*, 2621–2625.
33. Kapur, V.; Bansal, A.; Le, P.; Asensio, O. Non-vacuum processing of CuIn$_{1-x}$Ga$_x$Se$_2$ solar cells on rigid and flexible substrates using nanoparticle precursor inks. *Thin Solid Films* **2003**, *431–432*, 53–57. [CrossRef]
34. Yeh, M.; Hsu, H.; Wang, K.; Hoa, S.; Chen, G.; Chen, H. Toward low-cost large-area CIGS thin film: Compositional and structural variations in sequentially electrodeposited CIGS thin films. *Sol. Energy* **2016**, *125*, 415–425. [CrossRef]
35. Kaelin, M.; Rudmanna, D.; Kurdesaua, F.; Meyerb, T.; Zogga, H.; Tiwaria, A.N. CIS and CIGS layers from selenized nanoparticle precursors. *Thin Solid Films* **2003**, *431–432*, 58–62. [CrossRef]
36. Roux, F.; Amtablian, S.; Anton, M.; Besnard, G.; Bilhaut, L.; Bommersbach, P.; Braillon, J.; Cayron, C.; Disdier, A.; Fournier, H.; et al. Chalcopyrite thin-film solar cells by industry-compatible ink-based process. *Sol. Energy Mater. Sol. Cells* **2013**, *115*, 86–92. [CrossRef]
37. Woo, H.-J.; Lee, W.-J.; Koh, E.-K.; Jang, S.I.; Kim, S.; Moon, H.; Kwon, S.-H. Plasma-Enhanced Atomic Layer Deposition of TiN Thin Films as an Effective Se Diffusion Barrier for CIGS Solar Cells. *Nanomaterials* **2021**, *11*, 370. [CrossRef] [PubMed]
38. Sun, Y.; Lin, S.; Li, W.; Cheng, S.; Zhang, Y.; Liu, Y.; Liu, W. Review on Alkali Element Doping in Cu(In,Ga)Se$_2$ Thin Films and Solar Cells. *Engineering* **2017**, *4*, 452–459. [CrossRef]
39. Pianezzi, F.; Bloesch, P.; Chirilă, A.; Seyrling, S.; Buecheler, S.; Kranz, L.; Fella, C.; Tiwari, A.N. Electronic properties of Cu(In,Ga)Se$_2$ solar cells on stainless steel foils without diffusion barrier. *Prog. Photovolt. Res. Appl.* **2012**, *20*, 253–259. [CrossRef]
40. Wuerz, R.; Eicke, A.; Frankenfeld, M.; Kessler, F.; Powalla, M.; Rogin, P.; Yazdani-Assl, O. CIGS thin-film solar cells on steel substrates. *Thin Solid Films* **2009**, *517*, 2415–2418. [CrossRef]
41. Eisenbarth, T.; Caballero, R.; Kaufmann, C.A.; Eicke, A.; Unold, T. Influence of iron on defect concentrations and device performance for Cu(In,Ga)Se$_2$ solar cells on stainless steel substrates. *Prog. Photovolt. Res. Appl.* **2012**, *20*, 568–574. [CrossRef]

42. Jackson, P.; Grabitz, P.; Strohm, A.; Bilger, G.; Schock, H.W. Contamination of Cu(In,Ga)Se$_2$ by Metallic Substrates. In Proceedings of the 19th European Photovoltaic Solar Energy Conference and Exhibition, Paris, France, 7–11 June 2004; WIP-Munich: München, Germany; pp. 1936–1938.
43. Pianezzi, F.; Nishiwaki, S.; Kranz, L.; Sutter-Fella, C.M.; Reinhard, P.; Bissig, B.; Hagendorfer, H.; Buecheler, S.; Tiwari, A.N. Influence of Ni and Cr impurities on the electronic properties of Cu(In,Ga)Se$_2$ thin film solar cells. *Prog. Photovolt. Res. Appl.* **2015**, *23*, 892–900. [CrossRef]
44. Escorihuela, J.; García-Bernabé, A.; Montero, A.; Sahuquillo, O.; Giménez, E.; Compañ, V. Ionic Liquid Composite Polybenzimidazol Membranes for High Temperature PEMFC Applications. *Polymers* **2019**, *11*, 732. [CrossRef] [PubMed]
45. Escorihuela, J.; García-Bernabé, A.; Montero, A.; Andrio, A.; Sahuquillo, O.; Giménez, E.; Compañ, V. Proton Conductivity through Polybenzimidazole Composite Membranes Containing Silica Nanofiber Mats. *Polymers* **2019**, *11*, 1182. [CrossRef] [PubMed]
46. Escorihuela, J.; García-Bernabé, A.; Compañ, V. A Deep Insight into Different Acidic Additives as Doping Agents for Enhancing Proton Conductivity on Polybenzimidazole Membranes. *Polymers* **2020**, *12*, 1374. [CrossRef] [PubMed]
47. Fuentes, I.; Andrio, A.; García-Bernabé, A.; Escorihuela, J.; Viñas, C.; Teixidor, F.; Compañ, V. Structural and dielectric properties of Cobaltacarborane Composite Polybenzimidazole Membranes as solid polymer electrolytes at high temperature. *Phys. Chem. Chem. Phys.* **2018**, *20*, 10173–10184. [CrossRef] [PubMed]
48. Compañ, V.; Escorihuela, J.; Olvera, J.; García-Bernabé, A.; Andrio, A. Influence of the anion on diffusivity and mobility of ionic liquids composite polybenzimidazol membranes. *Electrochim. Acta* **2020**, *354*, 136666. [CrossRef]
49. Olvera-Mancilla, J.; Escorihuela, J.; Alexandrova, L.; Andrio, A.; García-Bernabé, A.; del Castillo, L.F.; Compañ, V. Effect of metallacarborane salt H[COSANE] doping on the performance properties of polybenzimidazole membranes for high temperature PEMFCs. *Soft Matter* **2020**, *16*, 7624–7635. [CrossRef] [PubMed]
50. Escorihuela, J.; Sahuquillo, Ó.; García-Bernabé, A.; Giménez, E.; Compañ, V. Phosphoric Acid Doped Polybenzimidazole (PBI)/Zeolitic Imidazolate Framework Composite Membranes with Significantly Enhanced Proton Conductivity under Low Humidity Conditions. *Nanomaterials* **2018**, *8*, 775. [CrossRef]
51. Kim, Y.-I.; Kim, K.-B.; Kim, M. Characterization of lattice parameters gradient of Cu(In$_{1-x}$Ga$_x$)Se$_2$ absorbing layer in thin-film solar cell by glancing incidence X-ray diffraction technique. *J. Mater. Sci. Technol.* **2020**, *51*, 193–201. [CrossRef]
52. Sim, J.-K.; Lee, S.-K.; Kim, J.-S.; Jeong, K.-U.; Ahn, H.-K.; Lee, C.-R. Efficiency enhancement of CIGS compound solar cell fabricated using homomorphic thin Cr$_2$O$_3$ diffusion barrier formed on stainless steel substrate. *Appl. Surf. Sci.* **2016**, *389*, 645–650. [CrossRef]
53. Thirumoorthi, M.; Prakash, J.T.J. Structure, optical and electrical properties of indium tin oxide ultra thin films prepared by jet nebulizer spray pyrolysis technique. *J. Asian Ceram. Soc.* **2016**, *4*, 124–132. [CrossRef]
54. Zhao, D.; Wu, Y.; Tu, B.; Xing, G.; Li, H.; He, Z. Understanding the Impact of Cu-In-Ga-S Nanoparticles Compactness on Holes Transfer of Perovskite Solar Cells. *Nanomaterials* **2019**, *9*, 286. [CrossRef] [PubMed]
55. Gariano, G.; Lesnyak, V.; Brescia, R.; Bertoni, G.; Dang, Z.; Gaspari, R.; De Trizio, L.; Manna, L. Role of the Crystal Structure in Cation Exchange Reactions Involving Colloidal Cu$_2$Se Nanocrystals. *J. Am. Chem. Soc.* **2017**, *139*, 9583–9590. [CrossRef] [PubMed]
56. Li, J.; Jin, Z.; Liu, T.; Wang, J.; Wang, D.; Lai, J.; Du, H.; Cui, L. Ternary and quaternary chalcopyrite Cu(In$_{1-x}$Ga$_x$)Se$_2$ nanocrystals: Organoalkali-assisted diethylene glycol solution synthesis and band-gap tuning. *CrystEngComm* **2013**, *15*, 7327–7338. [CrossRef]
57. Makuła, P.; Pacia, M.; Macyk, W. How To Correctly Determine the Band Gap Energy of Modified Semiconductor Photocatalysts Based on UV–Vis Spectra. *J. Phys. Chem. Lett.* **2018**, *9*, 6814–6817. [CrossRef]
58. del Castillo, R.M.; del Castillo, L.F.; Callesa, A.G.; Compañ, V. Experimental and computational conductivity study of multilayer graphene in polypropylene nanocomposites. *J. Mater. Chem. C* **2018**, *6*, 7232–7241. [CrossRef]
59. Coelho, R. Sur la relaxation d'une charge d'espace. *Revue Phys. Appl.* **1983**, *18*, 137. [CrossRef]
60. MacDonald, J.R. Theory of ac Space-Charge Polarization Effects in Photoconductors, Semiconductors, and Electrolytes. *Phys. Rev.* **1953**, *92*, 4. [CrossRef]
61. Sangoro, J.R.; Serghei, A.; Naumov, S.; Galvosas, P.; Kärger, J.; Wespe, C.; Bordusa, F.; Kremer, F. Charge Transport and Mass Transport in Imidazolium-Based Ionic Liquids. *Phys. Rev. E* **2008**, *77*, 051202. [CrossRef]
62. Serguei, A.; Tress, M.; Sangoro, J.R.; Kremer, F. Electrode Polarization and Charge Transport at Solid Interfaces. *Phys. Rev. B* **2009**, *80*, 184301. [CrossRef]
63. Jonscher, A.K. The 'Universal' Dielectric Response. *Nature* **1977**, *267*, 673–679. [CrossRef]
64. Jonscher, A.K. *Dielectric Relaxation in Solids*; Chelsea Dielectric Press Limited: London, UK, 1983.
65. Leys, J.; Wübbenhorst, M.; Preethy Menon, C.; Rajesh, R.; Thoen, J.; Glorieux, C.; Nockemann, P.; Thijs, B.; Binnemans, K.; Longuemart, S. Temperature dependence of the electrical conductivity of imidazolium ionic liquids. *J. Chem Phys.* **2008**, *128*, 064509. [CrossRef]
66. Greenhoe, B.M.; Hassan, M.K.; Wiggins, J.S.; Mauritz, K.A. Universal power law behavior of the AC conductivity versus frequency of agglomerate morphologies in conductive carbon nanotube-reinforced epoxy networks. *J. Polym. Sci. B* **2016**, *54*, 1918–1923. [CrossRef]
67. Haile, S.M.; Lentz, G.; Kreuer, K.-D.; Maier, J. Superprotonic conductivity in Cs$_3$(HSO$_4$)$_2$(H$_2$PO$_4$). *Solid State Ion.* **1995**, *77*, 128–134. [CrossRef]

Article

CuCrO₂ Nanoparticles Incorporated into PTAA as a Hole Transport Layer for 85 °C and Light Stabilities in Perovskite Solar Cells

Bumjin Gil [1,†], Jinhyun Kim [1,†], Alan Jiwan Yun [1], Kimin Park [1], Jaemin Cho [1], Minjun Park [2] and Byungwoo Park [1,*]

1 Department of Materials Science and Engineering, Research Institute of Advanced Materials, Seoul National University, Seoul 08826, Korea; bestgil123@snu.ac.kr (B.G.); kim767@snu.ac.kr (J.K.); hangyeolee@snu.ac.kr (A.J.Y.); flamethrow@snu.ac.kr (K.P.); jjm7004@snu.ac.kr (J.C.)
2 Department of Chemical Engineering, Ulsan National Institute of Science and Technology, Ulsan 44919, Korea; sia835@unist.ac.kr
* Correspondence: byungwoo@snu.ac.kr
† These authors contributed equally to this work.

Received: 25 July 2020; Accepted: 21 August 2020; Published: 26 August 2020

Abstract: High-mobility inorganic CuCrO₂ nanoparticles are co-utilized with conventional poly(bis(4-phenyl)(2,5,6-trimethylphenyl)amine) (PTAA) as a hole transport layer (HTL) for perovskite solar cells to improve device performance and long-term stability. Even though CuCrO₂ nanoparticles can be readily synthesized by hydrothermal reaction, it is difficult to form a uniform HTL with CuCrO₂ alone due to the severe agglomeration of nanoparticles. Herein, both CuCrO₂ nanoparticles and PTAA are sequentially deposited on perovskite by a simple spin-coating process, forming uniform HTL with excellent coverage. Due to the presence of high-mobility CuCrO₂ nanoparticles, CuCrO₂/PTAA HTL demonstrates better carrier extraction and transport. A reduction in trap density is also observed by trap-filled limited voltages and capacitance analyses. Incorporation of stable CuCrO₂ also contributes to the improved device stability under heat and light. Encapsulated perovskite solar cells with CuCrO₂/PTAA HTL retain their efficiency over 90% after ~900-h storage in 85 °C/85% relative humidity and under continuous 1-sun illumination at maximum-power point.

Keywords: perovskite solar cell; hole transport layer; CuCrO₂ nanoparticles; thermal stability; light stability

1. Introduction

In the field of next-generation photovoltaics, organic-inorganic hybrid halide perovskite solar cells have gathered tremendous attention since their emergence due to their rapidly growing power conversion efficiency (PCE), micrometer-scale carrier diffusion length, high absorption coefficient over solar spectrum regions, small exciton binding energy, etc. [1–11]. However, its relatively poor stability is still a main bottleneck toward commercialization, which becomes more serious at elevated temperatures or under constant illumination due to the rapid degradation of materials along with the accelerated formation and migration of defects [12–18]. One of the most vulnerable components is traditional organic small-molecule-based hole transport layers (HTL) such as 2,2′,7,7′-tetrakis(N,N-di-p-methoxyphenylamine)-9,9′-spirobifuorene (spiro-OMeTAD), which can easily decompose under the presence of heat [19,20]. Other candidates, such as poly(bis(4-phenyl)(2,5,6-trimethylphenyl)amine) (PTAA), are reported to be more durable in terms of stability [21–25], but diffusion of additives and ionic species can still occur to hamper the perovskite-HTL interface [26–29].

As an alternative to the unstable organic HTLs, inorganic HTLs such as CuSCN and various metal oxides have been shown to achieve long-term stability [30–44]. Among them, delafossite metal oxide $CuCrO_2$ is considered as one of the most promising candidates as an HTL due to its high mobility of 0.1–1 $cm^2\ V^{-1}\ s^{-1}$, favorable band alignment with perovskite, and facile synthesis method of nanoparticles by hydrothermal reaction of nitrate-based precursors [45–51]. Several research groups have adopted $CuCrO_2$ HTL in a *p-i-n* structure to obtain ambient stability comparable to its organic counterparts [52–55]. However, few studies have utilized $CuCrO_2$ in an *n-i-p* structure, mainly due to its difficulty in forming a uniform film over the perovskite layer [56]. Studies demonstrating long-term stabilities under continuous heat or light are also lacking; thus, a lot of effort is still required to successfully utilize $CuCrO_2$ materials as a stable and efficient HTL.

One strategy to overcome the barrier of poor film formability of nanoparticle-type HTL is to co-utilize with other HTL that can form homogeneous precursor solutions, which can have multiple advantages over single-component solutions. The solution-based secondary HTL can successfully immerse between nanoparticles, which can greatly improve the film uniformity and thereby reduce surface/interface-related defects. The ability to utilize high-mobility nanoparticles can also improve the overall hole mobility of the HTL and the stability of the perovskite-HTL interface, especially when the solution-based HTL is known to be susceptible to the interfacial degradation. Several studies have demonstrated this hybrid-type design, such as NiO_x/spiro-OMeTAD, NiO_x/CuSCN, and $CuGaO_2$/CuSCN, indicating the potential for further improvement of HTL by this co-utilization approach [57–59].

In this work, hydrothermally synthesized $CuCrO_2$ nanoparticles are incorporated into the conventional PTAA to form $CuCrO_2$/PTAA hybrid HTL that can effectively reduce the surface roughness. The utilization of high-mobility and stable $CuCrO_2$ can boost the hole extraction while passivating deep-level traps, which are confirmed by optoelectronic analyses. Stabilities of ~900 h under 85 °C/85% relative humidity (RH) and continuous 1-sun illumination further confirm the successful durability of the bilayer HTL, suggesting a straightforward but effective method to improve the stabilities of perovskite solar cells.

2. Materials and Methods

2.1. Synthesis of $CuCrO_2$ Nanoparticles

$Cu(NO_3)_2 \cdot 2.5H_2O$ (Alfa Aesar, Heysham, UK) and $Cr(NO_3)_3 \cdot 9H_2O$ (Alfa Aesar, Heysham, UK) were dissolved in deionized water (DW) with concentration of 0.21 M each. After 15 min of stirring, 1.8 M of NaOH (Daejung, Siheung, Korea) was added, and the solution was stirred for another 15 min. Then, the solution was transferred to a Teflon-lined stainless-steel autoclave and placed in an oven with a temperature of 220 °C for 60 h. After the reaction, a dark-green precipitate containing $CuCrO_2$ nanoparticles was formed. The synthesized nanoparticles were centrifuged and sequentially washed with 1 N HCl (Daejung, Siheung, Korea) and isopropyl alcohol (IPA, Daejung, Siheung, Korea) four times, and stored in IPA for future use.

2.2. Device Fabrication

Glasses coated with indium-doped tin oxide (ITO) were cleaned in acetone (Daejung, Siheung, Korea), ethanol (Daejung, Siheung, Korea), and DW for 15 min each by sonication, followed by UV-ozone treatment for 15 min. For the electron transport layer, SnO_2 aqueous colloidal dispersion (Alfa Aesar, Heysham, UK) was diluted by DW to 2.5 wt. %, spin-coated on ITO at 3000 rpm for 30 s, and annealed at 120 °C for 30 min. Perovskite precursor solution of 1.3 M $Cs_{0.05}(FA_{0.85}MA_{0.15})_{0.95}Pb(I_{0.85}Br_{0.15})_3$ (FA and MA stand for formamidinium and methylammonium, respectively) was fabricated by dissolving PbI_2 (TCI, Fukaya, Japan), $PbBr_2$ (TCI, Fukaya, Japan), FAI (Greatcell Solar, Queanbeyan, Australia), MABr (Greatcell Solar, Queanbeyan, Australia), and CsI (TCI, Fukaya, Japan) with desired ratio in a 4:1 (*v/v*) mixture of N,N-dimethylformamide (DMF, Sigma-Aldrich, St. Louis, MO, USA) and dimethyl sulfoxide (DMSO, Sigma-Aldrich, St. Louis, MO, USA). In a N_2-filled glovebox, the perovskite solution was deposited on a SnO_2 layer by spin-coating at 1000 rpm for 10 s, followed by 5000 rpm for 20 s. A total of 300 µL of chlorobenzene (Sigma-Aldrich, St. Louis, MO, USA) was dripped onto the spinning substrate 3 s before the end of the spin-coating process. The samples were then annealed at 100 °C for 40 min. For the $CuCrO_2$ hole transport layer, the stored $CuCrO_2$ nanoparticle dispersion was further diluted by IPA to the desired concentrations (0.5–3 mg mL^{-1}), subjected to sonication for 1 h, and spin-coated at 5000 rpm for 30 s, followed by annealing at 50 °C for 10 min to remove residual solvent. For a $CuCrO_2$-only device, the spin-coating steps were repeated multiple times to obtain full coverage of HTL, whereas for a $CuCrO_2$/PTAA device, single spin-coating of $CuCrO_2$ was sufficient. For the PTAA hole transport layer, solution was fabricated by dissolving 20 mg of PTAA (47 kDa, MS Solutions, Seoul, Korea) in 1 mL of chlorobenzene, with the addition of 6 µL of 4-*tert*-butylpyridine (Sigma-Aldrich, St. Louis, MO, USA) and 4 µL of 520 mg mL^{-1} bis(trifluoromethane)sulfonimide lithium salt (Sigma-Aldrich, St. Louis, MO, USA) solution in acetonitrile (Sigma-Aldrich, St. Louis, MO, USA). The PTAA solution was then spin-coated on either perovskite film or pre-deposited $CuCrO_2$ film at 3000 rpm for 30 s. Finally, an Au electrode was deposited by thermal evaporation. For encapsulated devices, the devices were sealed with cover glass using UV-curable epoxy resin (Nagase, Osaka, Japan).

2.3. Characterization

X-ray diffraction was conducted using a diffractometer (New D8 Advance, Bruker, Billerica, MA, USA). Surface roughness of the film was analyzed by an atomic force microscope (NX-10, Park Systems, Suwon, Korea). The cross-sectional image of an HTL film was obtained by a field-emission scanning electron microscope (Merlin-Compact, Zeiss, Oberkochen, Germany). The optical bandgap was analyzed by UV-visible absorption spectroscopy using a spectrophotometer (V-770, JASCO, Easton, MD, USA). A time-of-flight secondary ion mass spectrometer (TOF-SIMS-5, IONTOF, Münster, Germany) was utilized to obtain the depth profile of the device. Photoluminescence (LabRam HV Evolution, Horiba, Kyoto, Japan) and time-resolved photoluminescence (FluoTime 300, Picoquant, Berlin, Germany) of the films were analyzed using lasers with excitation wavelengths of 523 nm and 398 nm, respectively. Space-charge limited current (SCLC) and admittance analyses were conducted using a potentiostat (Zive SP-1, WonATech, Seoul, Korea), where dark current was measured under varying direct current (DC) bias for SCLC measurement and impedance was measured at a frequency of 10^{-2}–10^4 Hz using 10 mV AC voltage perturbation for admittance analysis. J-V curves of the solar cells were obtained using a solar simulator (K3000, McScience, Suwon, Korea) with 1-sun (AM 1.5G) illumination on the glass/ITO side, with a voltage sweep between 1.2 and −0.1 V, a scan rate of 100 mV s^{-1}, and active areas for solar cells were 0.09 cm^2. For the thermal stability test, encapsulated devices were stored within a dark test chamber (TH-PE-025, JeioTech, Daejeon, Korea) with controlled temperature and humidity (85 °C/85% RH), and J-V scans of devices were periodically conducted. For the light stability test, encapsulated devices were tested with maximum-power-point tracking equipment (K3600, McScience, Suwon, Korea) under continuous 1-sun illumination, where maximum-power voltage is constantly applied to the cells during the test.

3. Results and Discussion

One of the prerequisites for HTL in an *n-i-p* type perovskite solar cell is continuous film formability that can yield a thin and compact layer above a perovskite substrate. The $CuCrO_2$/PTAA HTL deposited on the conventional triple-cation perovskite ($Cs_{0.05}(FA_{0.85}MA_{0.15})_{0.95}Pb(I_{0.85}Br_{0.15})_3$) displayed a smooth surface topography with reasonably small surface roughness (Figure 1a). However, due to the agglomeration of $CuCrO_2$ nanoparticles when deposited on perovskite substrate, simple one-step spin-coating of $CuCrO_2$ nanoparticles alone often yielded incomplete coverage; hence, multiple spin-coatings were required for $CuCrO_2$ to fully cover the underlying perovskite layer (Figure S1a,b). Moreover, even though the full coverage was obtained with only $CuCrO_2$, the resulting HTL displayed a much larger surface roughness compared to the $CuCrO_2$/PTAA, resulting in a low PCE (Figure 1b and Figure S1c). The difficulty in creating a uniform film with nanoparticles alone suggests that incorporating a small number of nanoparticles within other solution-processable HTL is more suitable to utilize high-mobility nanoparticles, as suggested in this work. As shown in the cross-sectional image in Figure 1c, the final CuCrO$_2$/PTAA bilayer showed compact and dense morphology with ~100 nm thickness, suggesting that the PTAA solution was effectively wetted and immersed among the CuCrO$_2$ nanoparticles, and thereby formed a uniform film without any visible structural imperfections. Further analyses by x-ray diffraction (XRD) revealed that hydrothermally synthesized CuCrO$_2$ nanoparticles consisted of a mixture of desirable rhombohedral and hexagonal delafossite phases without any detectable impurities (Figure S2a) [52], and the perovskite layer was not damaged or decomposed into impurities like PbI$_2$ after the deposition of CuCrO$_2$/PTAA HTL (Figure S2b).

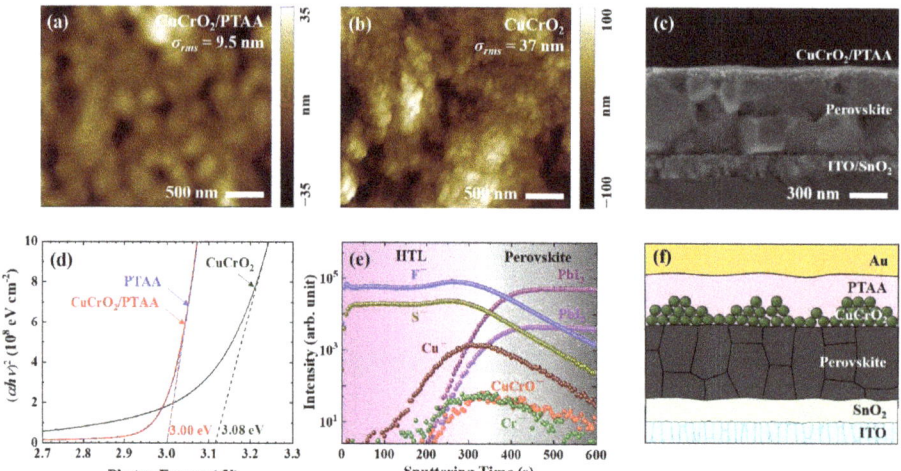

Figure 1. Morphological and structural analyses of CuCrO$_2$/PTAA and CuCrO$_2$ hole transport layer (HTL): (**a,b**) Topography and root-mean-square (RMS) surface roughness of each HTL deposited on indium-doped tin oxide (ITO)/SnO$_2$/perovskite, obtained by atomic force microscope (AFM); (**c**) Cross-sectional scanning electron microscopy (SEM) image of ITO/SnO$_2$/perovskite/HTL film; (**d**) UV-visible absorption spectra and the optical bandgap energy of each HTL deposited on a glass substrate; (**e**) Time-of-flight secondary ion mass spectrometry (TOF-SIMS) depth profile of the ITO/SnO$_2$/perovskite/CuCrO$_2$/PTAA film; (**f**) Schematic illustration of the device architecture with CuCrO$_2$/PTAA HTL. Light is incident on an ITO side during the solar cell operation.

The optical bandgap was determined to be 3.00 eV and 3.08 eV for PTAA-only and CuCrO$_2$-only HTL, respectively, as presented in Figure 1d [45,46,48–50,60,61]. The absorption spectrum and bandgap of the optimized CuCrO$_2$/PTAA layer were almost identical to those of PTAA, since a small number of CuCrO$_2$ nanoparticles was enough to form an effective and stable HTL layer. The presence and distribution of CuCrO$_2$ in the bilayer HTL was further confirmed by a time-of-flight secondary ion mass spectroscopy (TOF-SIMS), as shown in Figure 1e. PbI$_2^-$ and PbI$_3^-$ originated from the perovskite, and F$^-$ and S$^-$ originated from the additives of PTAA. Since species containing Cu and Cr were distributed near the perovskite-HTL interface, the CuCrO$_2$ nanoparticles were mainly located at the bottom part of the HTL, where they were percolated by the solution-processed PTAA (Figure 1f).

The electronic properties of CuCrO$_2$/PTAA hybrid HTL were investigated to evaluate its ability to extract and transport holes. Photoluminescence (PL) spectra in Figure 2a show that CuCrO$_2$/PTAA HTL exhibited slightly larger PL quenching compared to the bare PTAA, implying the increased hole-extracting ability due to the incorporation of high-mobility CuCrO$_2$ nanoparticles. Time-resolved PL spectra, as seen in Figure 2b, exhibited faster early-stage decay with CuCrO$_2$/PTAA compared to

PTAA, 14 vs. 28 ns, respectively [62–64]. To characterize the hole-extracting mobility more quantitatively, dark current-voltage (J-V) characteristics under DC bias were examined for the SCLC region with the ITO/HTL/Au structure (Figure 2c) [65,66]. The hole mobilities were 2.6×10^{-3} and 1.2×10^{-3} cm^2 V^{-1} s^{-1} for CuCrO$_2$/PTAA and bare PTAA, respectively, further supporting the role of high-mobility CuCrO$_2$ nanoparticles which enable faster hole extraction from the perovskite along the HTL.

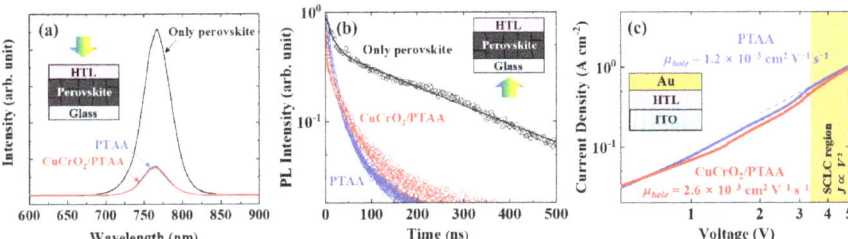

Figure 2. Electronic properties of CuCrO$_2$/PTAA HTL: (**a**) Steady-state photoluminescence (PL) and (**b**) time-resolved PL spectra (398-nm excitation with fitting lines) of bare perovskite and perovskite/HTL films deposited on glass; (**c**) Dark J-V characteristics of ITO/HTL/Au layers with different HTLs, where space-charge limited current (SCLC) region is indicated.

Next, the effect of CuCrO$_2$ incorporation into PTAA on the defect characteristics of the devices is discussed. Figure 3a shows dark J-V characteristics of hole-only devices with the structure of ITO/PTAA/perovskite/HTL/Au, with the upper HTL being either PTAA or CuCrO$_2$/PTAA. The trap-filled limited voltages (V_{TFL}) are related to the trap densities of the devices (N_t^{TFL}), exhibiting 6.6×10^{15} and 1.1×10^{16} cm^{-3} for the CuCrO$_2$/PTAA and bare PTAA, respectively [67]. Defect densities were also analyzed by capacitance analyses on the conventional ITO/SnO$_2$/perovskite/HTL/Au solar-cell structures. Nyquist plots in Figure 3b show larger semicircles for the device with CuCrO$_2$/PTAA compared to the PTAA, implying the increased recombination resistance which may be related to the decreased trap sites along the perovskite-HTL region [68–71]. It can also be seen in Figure 3c that two devices show different capacitive responses at low frequencies, indicating differences in the midgap trap states [68,72,73]. The trap density of states derived from the derivative of the capacitance (Figure 3d) exhibited a lower density of states in CuCrO$_2$/PTAA HTL, resulting in an almost halved integrated trap density (N_t^C) compared to the PTAA HTL [74–76]. These combined results of trap reduction suggest that CuCrO$_2$ nanoparticles near the perovskite-HTL interface surely passivate defects and related trap states, which can also contribute to the improvement of charge transport, as previously mentioned.

Solar cells with the device structure of ITO/SnO$_2$/perovskite/HTL/Au were fabricated with either CuCrO$_2$/PTAA or PTAA. With the optimum concentration of CuCrO$_2$ nanoparticles (Figure S3a), the champion cell yielded V_{OC} = 1.02 V, J_{SC} = 22.8 mA cm^{-2} and FF = 0.75 (PCE of 17.4%), whereas the bare PTAA yielded V_{OC} = 1.02 V, J_{SC} = 22.4 mA cm^{-2} and FF = 0.74 (PCE of 16.9%), as shown in Figure 4a (with the device parameters for multiple cells presented in Table 1 and Figure S3). While the external quantum efficiencies (EQEs) of the devices were quite similar (Figure 4c), the response at a longer wavelength (~700 nm or ~1.7 eV) exhibited better efficiency with the CuCrO$_2$-nanoparticle device, indicating an improved hole carrier collection, consistent with Figures 2 and 3. This improved hole collectivity might contribute to the increase of average J_{SC}, whereas the slight increases in trap-dependent parameters such as V_{OC} and FF further confirm the defect passivation effect by CuCrO$_2$ nanoparticles at the perovskite-HTL interface [77,78].

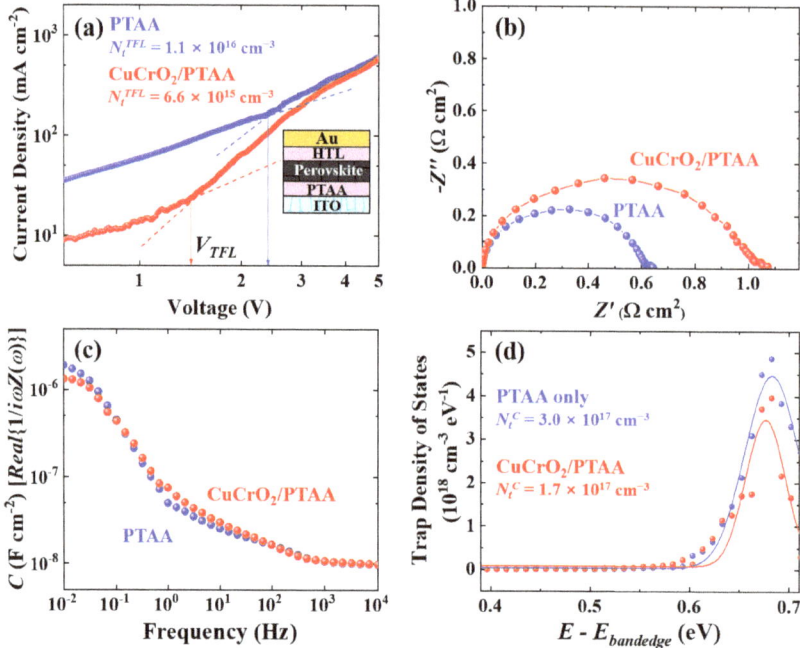

Figure 3. Trap density analyses of devices with different HTLs: (**a**) Dark J-V characteristics of hole-only devices with different upper HTLs, and calculated trap densities (N_t^{TFL}) from trap-filled limited voltages (V_{TFL}); (**b**) Nyquist plot, (**c**) capacitance-frequency plot, and (**d**) trap density of states obtained from the capacitances (with the integrated trap density N_t^C), in the device structure of ITO/SnO$_2$/perovskite/HTL/Au.

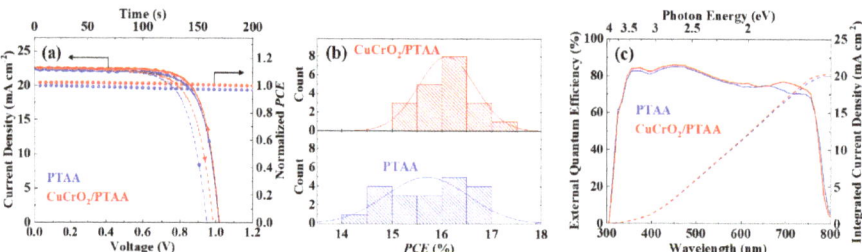

Figure 4. Photovoltaic performances of perovskite solar cells with CuCrO$_2$/PTAA or PTAA as HTL: (**a**) J-V curves of champion cells and their steady-state efficiencies under maximum power voltage; (**b**) PCE distributions for 20 cells at each condition; (**c**) External quantum efficiency (EQE) of solar cells.

Table 1. Photovoltaic parameters of the solar cells (reverse scan for 20 cells). The data in parentheses are from the cells with the best power conversion efficiency (PCE).

HTL	V_{OC} (V)	J_{SC} (mA cm^{-2})	FF	PCE (%)	HI $(1-\eta_{FOR}/\eta_{REV})$ [1]
PTAA	1.02 ± 0.03 (1.02)	21.3 ± 1.1 (22.4)	0.72 ± 0.02 (0.74)	15.7 ± 0.8 (16.9)	0.09 ± 0.04
CuCrO$_2$/PTAA	1.03 ± 0.15 (1.02)	21.6 ± 3.3 (22.8)	0.73 ± 0.11 (0.75)	16.1 ± 2.4 (17.4)	0.11 ± 0.04

[1] HI, η_{FOR} and η_{REV} refer to the hysteresis index, forward-scan PCE and reverse-scan PCE, respectively.

The effect of CuCrO$_2$ nanoparticles on both thermal and light stabilities of the solar cells were also investigated. For thermal stability, encapsulated devices were stored under standard damp heat conditions (85 °C/85% relative humidity (RH)) [25,79,80], where encapsulation was applied to block other external degradation factors than heat. High humidity was used to detect devices with damaged encapsulation which would undergo rapid moisture-induced degradation with a leak. As presented in Figure 5, improved thermal stability was observed for the device with CuCrO$_2$/PTAA HTL, where the device maintained over 90% of its initial PCE after 860 h. The degradation of organic cations in the perovskite or small organic molecules within HTL can critically damage both bulk and the interface, especially at an elevated temperature [9,10,20,81,82]. It can be inferred that the presence of more heat-resistant CuCrO$_2$ nanoparticles in the vicinity of perovskite and HTL creates a more heat-resistant interface with reduced interfacial reactions to maintain excellent thermal stability.

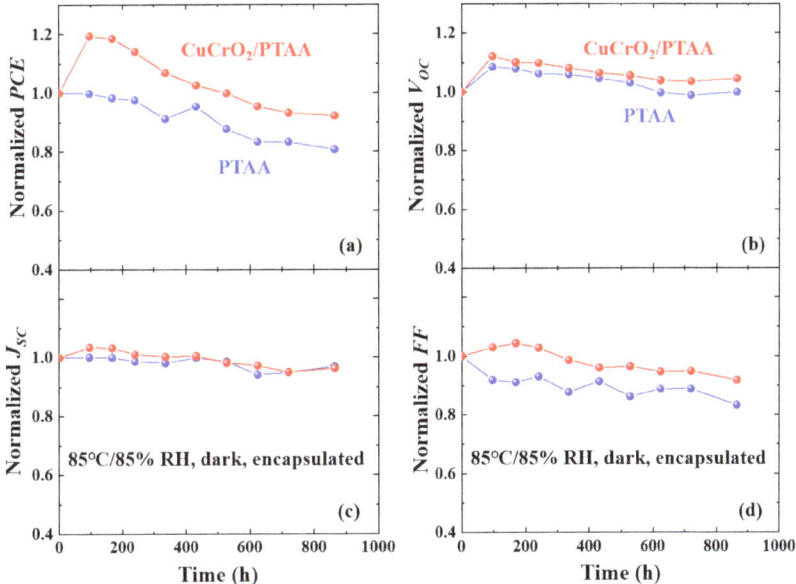

Figure 5. Thermal stabilities of solar cells with CuCrO$_2$/PTAA or PTAA HTL: Normalized values of (**a**) efficiency, (**b**) V$_{OC}$, (**c**) J$_{SC}$, and (**d**) FF of the encapsulated solar cells stored under 85 °C/85% relative humidity (RH) dark condition.

Light stabilities were tested by maximum-power-point tracking (MPPT) under continuous 1-sun (AM 1.5G) illumination. Figure 6 shows that a solar cell adopting CuCrO$_2$/PTAA HTL retains almost the entirety of its initial PCE after 960 h of operation, demonstrating superior light stability over the bare-PTAA device. Migration of halide defects as well as Li$^+$ from the additive in PTAA can accumulate at the perovskite-HTL interface and trigger interfacial degradation under operation conditions [15,28,29,83]. The improved light stability confirms that CuCrO$_2$ nanoparticles can directly prevent potential accumulation of traps at the interface, resulting in superior solar cells.

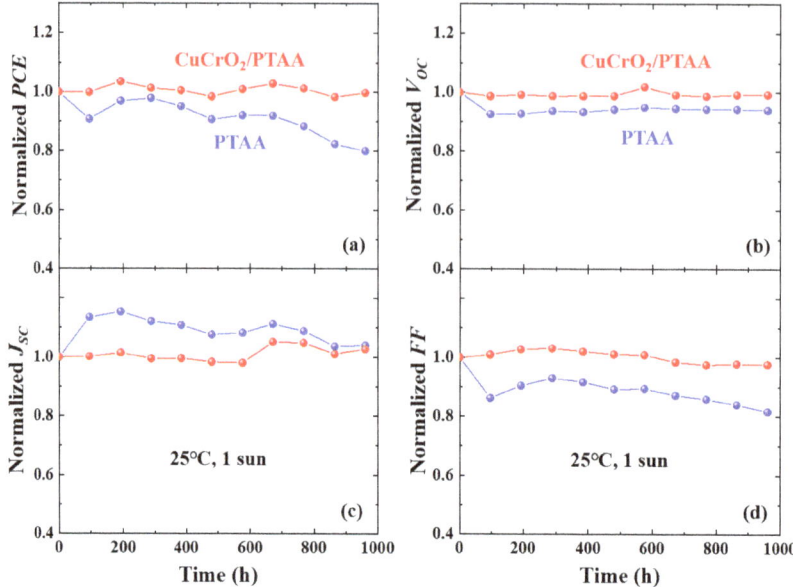

Figure 6. Light stabilities of solar cells with CuCrO$_2$/PTAA or PTAA HTL: Normalized values of (**a**) efficiency, (**b**) V$_{OC}$, (**c**) J$_{SC}$, and (**d**) FF of the encapsulated solar cells obtained by maximum-power-point tracking (MPPT) under continuous illumination of AM 1.5G at 25 °C.

4. Conclusions

Hybrid HTL, consisting of high-mobility CuCrO$_2$ nanoparticles embedded between perovskite and PTAA, was facilely adopted in the perovskite solar cells to guarantee excellent thermal and light stabilities. By a simple solution process, CuCrO$_2$/PTAA HTL was fabricated, yielding a uniform and smooth morphology. With CuCrO$_2$ nanoparticles providing high-mobility charge transport paths, CuCrO$_2$/PTAA HTL demonstrated more efficient hole-extraction abilities than the bare PTAA, and trap density was reduced by nearly half with CuCrO$_2$ nanoparticles. Therefore, solar cells with bilayer CuCrO$_2$/PTAA yielded higher PCE than the conventional PTAA-based ones, and also maintained over 90% of the initial efficiencies after storage under 85 °C/85% RH or operating under 1-sun MPPT for ~900 h. Our novel design of organic-inorganic hybrid HTL can aid in developing perovskite-based devices with improved hole extractability and reduced defects/traps, which ultimately leads to the superior stabilities under thermally induced and light-induced conditions.

Supplementary Materials: The following are available online at http://www.mdpi.com/2079-4991/10/9/1669/s1. Figure S1: Solar cells adopting only CuCrO$_2$ as an HTL; Figure S2: X-ray diffraction of CuCrO$_2$ nanoparticles and perovskite/CuCrO$_2$/PTAA HTL; Figure S3: Performances of solar cells adopting either CuCrO$_2$/PTAA or PTAA as an HTL.

Author Contributions: Conceptualization, B.G., J.K., and B.P.; methodology, B.G. and J.K.; validation, A.J.Y., K.P., J.C., M.P., and B.P.; formal analysis, B.G., J.K., A.J.Y., and K.P.; investigation, B.G. and J.K.; resources, A.J.Y., K.P., J.C., and M.P.; data curation, B.G. and J.K.; writing—original draft preparation, B.G., J.K., and B.P.; writing—review and editing, A.J.Y., K.P., J.C., M.P., and B.P.; supervision, B.P.; project administration, B.P.; funding acquisition, B.P. All authors have read and agreed to the published version of the manuscript.

Funding: This research was funded by the Korea Institute of Energy Technology Evaluation and Planning (KETEP), grant number 20183010014470, and the National Research Foundation of Korea (NRF), grant number 2020R1A2C100545211.

Conflicts of Interest: The authors declare no conflict of interest.

References

1. Hodes, G. Perovskite-based solar cells. *Science* **2013**, *312*, 317–318. [CrossRef] [PubMed]
2. Snaith, H.J. Perovskites: The emergence of a new era for low-cost, high-efficiency solar cells. *J. Phys. Chem. Lett.* **2013**, *4*, 3623–3630. [CrossRef]
3. Stranks, S.D.; Eperon, G.E.; Grancini, G.; Menelaou, C.; Alcocer, M.J.P.; Leijtens, T.; Herz, L.M.; Petrozza, A.; Snaith, H.J. Electron-hole diffusion lengths exceeding 1 micrometer in an organometal trihalide perovskite absorber. *Science* **2013**, *342*, 341–344. [CrossRef]
4. Frost, J.M.; Butler, K.T.; Brivio, F.; Hendon, C.H.; Schilfgaarde, M.V.; Walsh, A. Atomistic origins of high-performance in hybrid halide perovskite solar cells. *Nano Lett.* **2014**, *14*, 2584–2590. [CrossRef] [PubMed]
5. Todorov, T.; Gershon, T.; Gunawan, O.; Lee, Y.S.; Sturdevant, S.; Chang, L.-Y.; Guha, S. Monolithic perovskite-CIGS tandem solar cells via in situ band gap engineering. *Adv. Energy Mater.* **2015**, *5*, 1500799. [CrossRef]
6. Park, N.-G. Perovskite solar cells: An emerging photovoltaic technology. *Mater. Today* **2015**, *18*, 65–72. [CrossRef]
7. Saliba, M.; Matsui, T.; Seo, J.-Y.; Domanski, K.; Correa-Baena, J.-P.; Nazeeruddin, M.K.; Zakeeruddin, S.M.; Tress, W.; Abate, A.; Hagfeldt, A.; et al. Cesium-containing triple cation perovskite solar cells: Improved stability, reproducibility and high efficiency. *Energy Environ. Sci.* **2016**, *9*, 1989–1997. [CrossRef]
8. Hwang, T.; Lee, B.; Kim, J.; Lee, S.; Gil, B.; Yun, A.J.; Park, B. From nanostructural evolution to dynamic interplay of constituents: Perspectives for perovskite solar cells. *Adv. Mater.* **2018**, *30*, 1704208. [CrossRef]
9. Kim, J.; Hwang, T.; Lee, B.; Lee, S.; Park, K.; Park, H.H.; Park, B. An aromatic diamine molecule as the *a*-site solute for highly durable and efficient perovskite solar cells. *Small Methods* **2019**, *3*, 1800361. [CrossRef]
10. Kim, J.; Yun, A.J.; Gil, B.; Lee, Y.; Park, B. Triamine-based aromatic cation as a novel stabilizer for efficient perovskite solar cells. *Adv. Funct. Mater.* **2019**, *29*, 1905190. [CrossRef]
11. Yang, A.; Blancon, J.-C.; Jiang, W.; Zhang, H.; Wong, J.; Yan, E.; Lin, Y.-R.; Crochet, J.; Kanatzidis, M.G.; Jariwala, D.; et al. Giant enhancement of photoluminescence emission in WS_2-two-dimensional perovskite heterostructures. *Nano Lett.* **2019**, *19*, 4852–4860. [CrossRef]
12. Fu, R.; Zhou, W.; Li, Q.; Zhao, Y.; Yu, D.; Zhao, Q. Stability challenges for perovskite solar cells. *ChemNanoMat* **2018**, *5*, 253–265. [CrossRef]
13. Gholipour, S.; Saliba, M. From exceptional properties to stability challenges of perovskite solar cells. *Small* **2018**, *14*, 1802385. [CrossRef] [PubMed]
14. Azpiroz, J.M.; Mosconi, E.; Bisquert, J.; De Angelis, F. Defect migration in methylammonium lead iodide and its role in perovskite solar cell operation. *Energy Environ. Sci.* **2015**, *8*, 2118–2127. [CrossRef]
15. Ruan, S.; Surmiak, M.-A.; Ruan, Y.; McMeekin, D.P.; Ebendorff-Heidepriem, H.; Cheng, Y.-B.; Lu, J.; McNeill, C.R. Light induced degradation in mixed-halide perovskites. *J. Mater. Chem. C* **2019**, *7*, 9326–9334. [CrossRef]
16. Holovský, J.; Amalathas, A.P.; Landová, L.; Dzurňák, B.; Conrad, B.; Ledinský, M.; Hájková, Z.; Pop-Georgievski, O.; Svoboda, J.; Yang, T.C.-J.; et al. Lead halide residue as a source of light-induced reversible defects in hybrid perovskite layers and solar cells. *ACS Energy Lett.* **2019**, *4*, 3011–3017. [CrossRef]
17. Park, C.-G.; Choi, W.-G.; Na, S.; Moon, T. All-inorganic perovskite $cspbi_2br$ through co-evaporation for planar heterojunction solar cells. *Electron. Mater. Lett.* **2019**, *15*, 56–60. [CrossRef]
18. Park, H.H.; Kim, J.; Kim, G.; Jung, H.; Kim, S.; Moon, C.S.; Lee, S.J.; Shin, S.S.; Hao, X.; Yun, J.S.; et al. Transparent electrodes consisting of a surface-treated buffer layer based on tungsten oxide for semitransparent perovskite solar cells and four-terminal tandem applications. *Small Methods* **2020**, *4*, 2070018. [CrossRef]
19. Malinauskas, T.; Tomkute-Luksiene, D.; Sens, R.; Daskeviciene, M.; Send, R.; Wonneberger, H.; Jankauskas, V.; Bruder, I.; Getautis, V. Enhancing thermal stability and lifetime of solid-state dye-sensitized solar cells via molecular engineering of the hole-transporting material spiro-OMeTAD. *ACS Appl. Mater. Interfaces* **2015**, *7*, 11107–11116. [CrossRef]
20. Jena, A.K.; Numata, Y.; Ikegamia, M.; Miyasaka, T. Role of spiro-OMeTAD in performance deterioration of perovskite solar cells at high temperature and reuse of the perovskite films to avoid Pb-Waste. *J. Mater. Chem. A* **2018**, *6*, 2219–2230. [CrossRef]
21. Ahn, N.; Jeon, I.; Yoon, J.; Kauppinen, E.I.; Matsuo, Y.; Maruyama, S.; Choi, M. Carbon-sandwiched perovskite solar cell. *J. Mater. Chem. A* **2018**, *6*, 1382–1389. [CrossRef]

22. Duong, T.; Wu, Y.; Shen, H.; Peng, J.; Wu, N.; White, T.; Weber, K.; Catchpole, K. Impact of light on the thermal stability of perovskite solar cells and development of stable semi-transparent cells. In Proceedings of the 2018 IEEE 7th World Conference on Photovoltaic Energy Conversion (WCPEC) (A Joint Conference of 45th IEEE PVSC, 28th PVSEC & 34th EU PVSEC), Waikoloa Village, HI, USA, 10–15 June 2018; pp. 3506–3508. [CrossRef]
23. Thote, A.; Jeon, I.; Lee, J.-W.; Seo, S.; Lin, H.-S.; Yang, Y.; Daiguji, H.; Maruyama, S.; Matsuo, Y. Stable and reproducible 2D/3D formamidinium-lead-iodide perovskite solar cells. *ACS Appl. Energy Mater.* **2019**, *2*, 2486–2493. [CrossRef]
24. Zhao, Q.; Wu, R.; Zhang, Z.; Xiong, J.; He, Z.; Fan, B.; Dai, Z.; Yang, B.; Xue, X.; Cai, P.; et al. Achieving efficient inverted planar perovskite solar cells with nondoped PTAA as a hole transport layer. *Org. Electron.* **2019**, *71*, 106–112. [CrossRef]
25. Matsui, T.; Yamamoto, T.; Nishihara, T.; Morisawa, R.; Yokoyama, T.; Sekiguchi, T.; Negami, T. Compositional engineering for thermally stable, highly efficient perovskite solar cells exceeding 20% power conversion efficiency with 85 °C/85% 1000 h stability. *Adv. Mater.* **2019**, *31*, 1806823. [CrossRef] [PubMed]
26. Berhe, T.A.; Su, W.-N.; Chen, C.-H.; Pan, C.-J.; Cheng, J.-H.; Chen, H.-M.; Tsai, M.-C.; Chen, L.-Y.; Dubale, A.A.; Hwang, B.-J. Organometal halide perovskite solar cells: Degradation and stability. *Energy Environ. Sci.* **2016**, *9*, 323–356. [CrossRef]
27. Kim, G.-W.; Kang, G.; Kim, J.; Lee, G.-Y.; Kim, H.I.; Pyeon, L.; Lee, J.; Park, T. Dopant-free polymeric hole transport materials for highly efficient and stable perovskite solar cells. *Energy Environ. Sci.* **2016**, *9*, 2326–2333. [CrossRef]
28. Yang, T.-Y.; Jeon, N.J.; Shin, H.-W.; Shin, S.S.; Kim, Y.Y.; Seo, J. achieving long-term operational stability of perovskite solar cells with a stabilized efficiency exceeding 20% after 1000 h. *Adv. Sci.* **2019**, *6*, 1900528. [CrossRef]
29. Schloemer, T.H.; Christians, J.A.; Luther, J.M.; Sellinger, A. Doping strategies for small molecule organic hole-transport materials: Impacts on perovskite solar cell performance and stability. *Chem. Sci.* **2019**, *10*, 1904–1935. [CrossRef]
30. Chen, J.; Park, N. Inorganic hole transporting materials for stable and high efficiency perovskite solar cells. *J. Phys. Chem. C* **2018**, *122*, 14039–14063. [CrossRef]
31. Gil, B.; Yun, A.J.; Lee, Y.; Kim, J.; Lee, B.; Park, B. Recent progress in inorganic hole transport materials for efficient and stable perovskite solar cells. *Electron. Mater. Lett.* **2019**, *15*, 505–524. [CrossRef]
32. Arora, N.; Dar, M.I.; Hinderhofer, A.; Pellet, N.; Schreiber, F.; Zakeeruddin, S.M.; Grätzel, M. Perovskite solar cells with CuSCN hole extraction layers yield stabilized efficiencies greater than 20%. *Science* **2017**, *358*, 768–771. [CrossRef] [PubMed]
33. Yue, S.; Liu, K.; Xu, R.; Li, M.; Azam, M.; Ren, K.; Liu, J.; Sun, Y.; Wang, Z.; Cao, D.; et al. Efficacious engineering on charge extraction for realizing highly efficient perovskite solar cells. *Energy Environ. Sci.* **2017**, *10*, 2570–2578. [CrossRef]
34. Wilson, S.S.; Bosco, J.P.; Tolstova, Y.; Scanlon, D.O.; Watson, G.W.; Atwater, H.A. Interface stoichiometry control to improve device voltage and modify band alignment in ZnO/Cu_2O heterojunction solar cells. *Energy Environ. Sci.* **2014**, *7*, 3606–3610. [CrossRef]
35. Lien, H.-T.; Wong, D.P.; Tsao, N.-H.; Huang, C.-I.; Su, C.; Chen, K.-H.; Chen, L.-C. Effect of copper oxide oxidation state on the polymer-based solar cell buffer layers. *ACS Appl. Mater. Interfaces* **2014**, *6*, 22445–22450. [CrossRef] [PubMed]
36. Rao, H.; Ye, S.; Sun, W.; Yan, W.; Li, Y.; Peng, H.; Liu, Z.; Bian, Z.; Li, Y.; Huang, C. A 19.0% efficiency achieved in CuO_x-based inverted $CH_3NH_3PbI_{3-x}Cl_x$ solar cells by an effective Cl doping method. *Nano Energy* **2016**, *27*, 51–57. [CrossRef]
37. Igbari, F.; Li, M.; Hu, Y.; Wang, Z.-K.; Liao, L.-S. A room temperature $CuAlO_2$ hole interfacial layer for efficient and stable planar perovskite solar cells. *J. Mater. Chem. A* **2016**, *4*, 1326–1335. [CrossRef]
38. Chen, Y.; Yang, Z.; Wang, S.; Zheng, X.; Wu, Y.; Yuan, N.; Zhang, W.-H.; Liu, S.F. Design of an inorganic mesoporous hole-transporting layer for highly efficient and stable inverted perovskite solar cells. *Adv. Mater.* **2018**, *30*, 1805660. [CrossRef]
39. Qin, P.; He, Q.; Yang, G.; Yua, X.; Xiong, L.; Fang, G. Metal ions diffusion at heterojunction chromium oxide/$CH_3NH_3PbI_3$ interface on the stability of perovskite solar cells. *Surf. Interfaces* **2018**, *10*, 93–99. [CrossRef]

40. Li, D.; Tong, C.; Ji, W.; Fu, Z.; Wan, Z.; Huang, Q.; Ming, Y.; Mei, A.; Hu, Y.; Rong, Y.; et al. Vanadium oxide post-treatment for enhanced photovoltage of printable perovskite solar cells. *ACS Sustain. Chem. Eng.* **2019**, *7*, 2619–2625. [CrossRef]
41. Shalan, A.E.; Oshikiri, T.; Narra, S.; Elshanawany, M.M.; Ueno, K.; Wu, H.-P.; Nakamura, K.; Shi, X.; Diau, E.W.-G.; Misawa, H. Cobalt oxide (CoO_x) as an efficient hole-extracting layer for high-performance inverted planar perovskite solar cells. *ACS Appl. Mater. Interfaces* **2016**, *8*, 33592–33600. [CrossRef]
42. Im, K.; Heo, J.H.; Im, S.H.; Kim, J.S. Scalable synthesis of Ti-doped MoO_2 nanoparticle-hole-transporting material with high moisture stability for $CH_3NH_3PbI_3$ perovskite solar cells. *Chem. Eng. J.* **2017**, *330*, 698–705. [CrossRef]
43. Lee, B.; Shin, B.; Park, B. Uniform Cs_2SnI_6 thin films for lead-free and stable perovskite optoelectronics via hybrid deposition approaches. *Electron. Mater. Lett.* **2019**, *15*, 192–200. [CrossRef]
44. Chen, W.-C.; Tunuguntla, V.; Chiu, M.-H.; Li, L.-J.; Shown, I.; Lee, C.-H.; Hwang, J.-S.; Chen, L.-C.; Chen, K.-H. Co-solvent effect on microwave-assisted Cu_2ZnSnS_4 nanoparticles synthesis for thin film solar cell. *Sol. Energy Mater. Sol. Cells* **2017**, *161*, 416–423. [CrossRef]
45. Li, D.; Fang, X.; Deng, Z.; Zhou, S.; Tao, R.; Dong, W.; Wang, T.; Zhao, Y.; Meng, G.; Zhu, X. Electrical, optical and structural properties of $CuCrO_2$ films prepared by pulsed laser deposition. *J. Phys. D Appl. Phys.* **2007**, *40*, 4910–4915. [CrossRef]
46. Wang, J.; Zheng, P.; Li, D.; Deng, Z.; Dong, W.; Tao, R.; Fang, X. Preparation of delafossite-type $CuCrO_2$ films by Sol–Gel method. *J. Alloys Compd.* **2011**, *509*, 5715–5719. [CrossRef]
47. Xiong, D.; Xu, Z.; Zeng, X.; Zhang, W.; Chen, W.; Xu, X.; Wang, M.; Cheng, Y.-B. Hydrothermal synthesis of ultrasmall $CuCrO_2$ nanocrystal alternatives to NiO nanoparticles in efficient p-Type dye-sensitized solar cells. *J. Am. Chem.* **2012**, *22*, 24760–24768. [CrossRef]
48. Yu, R.-S.; Wu, C.-M. Characteristics of p-Type transparent conductive $CuCrO_2$ thin films. *Appl. Surf. Sci.* **2013**, *282*, 92–97. [CrossRef]
49. Barnabé, A.; Thimont, Y.; Lalanne, M.; Presmanesa, L.; Tailhades, P. p-Type conducting transparent characteristics of delafossite Mg-doped $CuCrO_2$ thin films prepared by RF-Sputtering. *J. Mater. Chem. C* **2015**, *3*, 6012–6024. [CrossRef]
50. Sánchez-Alarcón, R.I.; Oropeza-Rosario, G.; Gutierrez-Villalobos, A.; Muro-López, M.A.; Martínez-Martínez, R.; Zaleta-Alejandre, E.; Falcony, C.; Alarcón-Flores, G.; Fragoso, E.; Hernández-Silva, O.; et al. Ultrasonic spray-pyrolyzed $CuCrO_2$ thin films. *J. Phys. D Appl. Phys.* **2016**, *49*, 175102. [CrossRef]
51. Nie, S.; Liu, A.; Meng, Y.; Shin, B.; Liu, G.; Shan, F. Solution processed ternary p-Type $CuCrO_2$ semiconductor thin films and their application in transistors. *J. Mater. Chem. C* **2018**, *6*, 1393–1398. [CrossRef]
52. Dunlap-Shohl, W.A.; Daunis, T.B.; Wang, X.; Wang, J.; Zhang, B.; Barrera, D.; Yan, Y.; Hsu, J.W.P.; Mitzi, D.B. Room-temperature fabrication of a delafossite $CuCrO_2$ hole transport layer for perovskite solar cells. *J. Mater. Chem. A* **2018**, *6*, 469–477. [CrossRef]
53. Zhang, H.; Wang, H.; Zhu, H.; Chueh, C.-C.; Chen, W.; Yang, S.; Jen, A.K.-Y. Low-temperature solution-processed $CuCrO_2$ hole-transporting layer for efficient and photostable perovskite solar cells. *Adv. Energy Mater.* **2018**, *8*, 1702762. [CrossRef]
54. Jeong, S.; Seo, S.; Shin, H. p-Type $CuCrO_2$ particulate films as the hole transporting layer for $CH_3NH_3PbI_3$ perovskite solar cells. *RSC Adv.* **2018**, *8*, 27956–27962. [CrossRef]
55. Yang, B.; Ouyang, D.; Huang, Z.; Ren, X.; Zhang, H.; Choy, W.C.H. Multifunctional synthesis approach of n: $CuCrO_2$ nanoparticles for hole transport layer in high-performance perovskite solar cells. *Adv. Funct. Mater.* **2019**, *29*, 1902600. [CrossRef]
56. Akin, S.; Liu, Y.; Dar, M.I.; Zakeeruddin, S.M.; Grätzel, M.; Turand, S.; Sonmezoglu, S. Hydrothermally processed $CuCrO_2$ nanoparticles as an inorganic hole transporting material for low cost perovskite solar cells with superior stability. *J. Mater. Chem. A* **2018**, *6*, 20327–20337. [CrossRef]
57. Cao, J.; Yu, H.; Zhou, S.; Qin, M.; Lau, T.-K.; Lu, X.; Zhao, N.; Wong, C.-P. Low-temperature solution-processed NiO_x films for air-stable perovskite solar cells. *J. Mater. Chem. A* **2017**, *5*, 11071–11077. [CrossRef]
58. Mali, S.S.; Patil, J.V.; Kim, H.; Luque, R.; Hong, C.K. Highly efficient thermally stable perovskite solar cells via $Cs:NiO_x$/CuSCN double-inorganic hole extraction layer interface engineering. *Mater. Today* **2019**, *26*, 8–18. [CrossRef]

59. Lee, B.; Yun, A.J.; Kim, J.; Gil, B.; Shin, B.; Park, B. Aminosilane-Modified CuGaO$_2$ Nanoparticles incorporated with CuSCN as a hole-transport layer for efficient and stable perovskite solar cells. *Adv. Mater. Interfaces* **2019**, *6*, 1901372. [CrossRef]
60. Panidi, J.; Paterson, A.F.; Khim, D.; Fei, Z.; Han, Y.; Tsetseris, L.; Vourlias, G.; Patsalas, P.A.; Heeney, M.; Anthopoulos, T.D. Remarkable enhancement of the hole mobility in several organic small-molecules, polymers, and small-molecule: Polymer blend transistors by simple admixing of the lewis acid p-dopant B(C$_6$F$_5$)$_3$. *Adv. Sci.* **2018**, *5*, 1700190. [CrossRef]
61. Pitchaiya, S.; Natarajan, M.; Santhanam, A.; Asokan, V.; Yuvapragasam, A.; Ramakrishnan, V.M.; Palanisamy, S.E.; Sundaram, S.; Velauthapillai, D. A review on the classification of organic/inorganic/carbonaceous hole transporting materials for perovskite solar cell application. *Arab. J. Chem.* **2020**, *13*, 2526–2557. [CrossRef]
62. Kim, J.I.; Kim, J.; Lee, J.; Jung, D.-R.; Kim, H.; Choi, H.; Lee, S.; Byun, S.; Kang, S.; Park, B. Photoluminescence enhancement in CdS quantum dots by thermal annealing. *Nanoscale Res. Lett.* **2012**, *7*, 482. [CrossRef]
63. Chen, J.; Kim, S.-G.; Ren, X.; Jung, H.S.; Park, N.-G. Effect of bidentate and tridentate additives on the photovoltaic performance and stability of perovskite solar cells. *J. Mater. Chem. A* **2019**, *7*, 4977–4987. [CrossRef]
64. Han, G.; Hadi, H.D.; Bruno, A.; Kulkarni, S.A.; Koh, T.M.; Wong, L.H.; Soci, C.; Mathews, N.; Zhang, S.; Mhaisalkar, S.G. Additive selection strategy for high performance perovskite photovoltaics. *J. Phys. Chem. C* **2018**, *122*, 13884–13893. [CrossRef]
65. Kim, Y.; Jung, E.H.; Kim, G.; Kim, D.; Kim, B.J.; Seo, J. Sequentially fluorinated PTAA polymers for enhancing V_{OC} of high-performance perovskite solar cells. *Adv. Energy Mater.* **2018**, *8*, 1801668. [CrossRef]
66. Kim, J.; Lee, Y.; Yun, A.J.; Gil, B.; Park, B. Interfacial modification and defect passivation by the cross-linking interlayer for efficient and stable CuSCN-based perovskite solar cell. *ACS Appl. Mater. Interfaces* **2019**, *11*, 46818–46824. [CrossRef] [PubMed]
67. Bube, R.H. Trap density determination by space-charge-limited currents. *J. Appl. Phys.* **1962**, *33*, 1733–1737. [CrossRef]
68. Guerrero, A.; Garcia-Belmonte, G.; Mora-Sero, I.; Bisquert, J.; Kang, Y.S.; Jacobsson, T.J.; Correa-Baena, J.-P.; Hagfeldt, A. Properties of contact and bulk impedances in hybrid lead halide perovskite solar cells including inductive loop elements. *J. Phys. Chem. C* **2016**, *120*, 8023–8032. [CrossRef]
69. Kim, J.; Hwang, T.; Lee, S.; Lee, B.; Kim, J.; Kim, J.; Gil, B.; Park, B. Synergetic effect of double-step blocking layer for the perovskite solar cell. *J. Appl. Phys.* **2017**, *122*, 145106. [CrossRef]
70. Lee, S.; Flanagan, J.C.; Lee, B.; Hwang, T.; Kim, J.; Gil, B.; Shim, M.; Park, B. Route to improving photovoltaics based on CdSe/CdSe$_x$Te$_{1-x}$ type-ii heterojunction nanorods: The effect of morphology and cosensitization on carrier recombination and transport. *ACS Appl. Mater. Interfaces* **2017**, *9*, 31931–31939. [CrossRef]
71. Lee, S.; Flanagan, J.C.; Kim, J.; Yun, A.J.; Lee, B.; Shim, M.; Park, B. Efficient type-ii heterojunction nanorod sensitized solar cells realized by controlled synthesis of core/patchy-shell structure and CdS cosensitization. *ACS Appl. Mater. Interfaces* **2019**, *11*, 19104–19114. [CrossRef]
72. Almora, O.; Aranda, C.; Mas-Marzá, E.; Garcia-Belmonte, G. On mott-schottky analysis interpretation of capacitance measurements in organometal perovskite solar cells. *Appl. Phys. Lett.* **2016**, *109*, 173903. [CrossRef]
73. Gunawan, O.; Gokmen, T.; Warren, C.W.; Cohen, J.D.; Todorov, T.K.; Barkhouse, D.A.R.; Bag, S.; Tang, J.; Shin, B.; Mitzi, D.B. Electronic properties of the Cu$_2$ZnSn(Se,S)$_4$ absorber layer in solar cells as revealed by admittance spectroscopy and related methods. *Appl. Phys. Lett.* **2012**, *100*, 253905. [CrossRef]
74. Walter, T.; Herberholz, R.; Müller, C.; Schock, H.W. Determination of defect distributions from admittance measurements and application to Cu(In,Ga)Se$_2$ based heterojunctions. *J. Appl. Phys.* **1996**, *80*, 4411–4420. [CrossRef]
75. Reese, M.O.; Gevorgyan, S.A.; Jørgensen, M.; Bundgaard, E.; Kurtz, S.R.; Ginley, D.S.; Olson, D.C.; Lloyd, M.T.; Moryillo, P.; Katz, E.A.; et al. Consensus stability testing protocols for organic photovoltaic materials and devices. *Sol. Energy Mater. Sol. Cells* **2011**, *95*, 1253–1267. [CrossRef]
76. Domanski, K.; Alharbi, E.A.; Hagfeldt, A.; Grätzel, M.; Tress, W. Systematic investigation of the impact of operation conditions on the degradation behaviour of perovskite solar cells. *Nat. Energy* **2018**, *3*, 61–67. [CrossRef]

77. Hwang, T.; Yun, A.J.; Kim, J.; Cho, D.; Kim, S.; Hong, S.; Park, B. Electronic traps and their correlations to perovskite solar cell performance via compositional and thermal annealing controls. *ACS Appl. Mater. Interfaces* **2019**, *11*, 6907–6917. [CrossRef]
78. Yun, A.J.; Kim, J.; Hwang, T.; Park, B. Origins of efficient perovskite solar cells with low-temperature processed SnO$_2$ electron transport layer. *ACS Appl. Energy Mater.* **2019**, *2*, 3554–3560. [CrossRef]
79. Ni, Z.; Bao, C.; Liu, Y.; Jiang, Q.; Wu, W.-Q.; Chen, S.; Dai, X.; Chen, B.; Hartweg, B.; Yu, Z.; et al. Resolving spatial and energetic distributions of trap states in metal halide perovskite solar cells. *Science* **2020**, *367*, 1352–1358. [CrossRef]
80. Hwang, T.; Yun, A.J.; Lee, B.; Kim, J.; Lee, Y.; Park, B. Methylammonium-chloride post-treatment on perovskite surface and its correlation to photovoltaic performance in the aspect of electronic traps. *J. Appl. Phys.* **2019**, *126*, 023101. [CrossRef]
81. Tan, W.; Bowring, A.R.; Meng, A.C.; McGehee, M.D.; McIntyre, P.C. Thermal stability of mixed cation metal halide perovskites in air. *ACS Appl. Mater. Interfaces* **2018**, *10*, 5485–5491. [CrossRef]
82. Szostak, R.; Silva, J.C.; Turren-Cruz, S.-H.; Soares, M.M.; Freitas, R.O.; Hagfeldt, A.; Tolentino, H.C.N.; Nogueira, A.F. Nanoscale mapping of chemical composition in organic-inorganic hybrid perovskite films. *Sci. Adv.* **2019**, *5*, eaaw6619. [CrossRef] [PubMed]
83. Seo, J.-Y.; Kim, H.-S.; Akin, S.; Stojanovic, M.; Simon, E.; Fleischer, M.; Hagfeldt, A.; Zakeeruddin, S.M.; Grätzel, M. Novel p-dopant toward highly efficient and stable perovskite solar cells. *Energy Environ. Sci.* **2018**, *11*, 2985–2992. [CrossRef]

© 2020 by the authors. Licensee MDPI, Basel, Switzerland. This article is an open access article distributed under the terms and conditions of the Creative Commons Attribution (CC BY) license (http://creativecommons.org/licenses/by/4.0/).

Article

On Tailoring Co-Precipitation Synthesis to Maximize Production Yield of Nanocrystalline Wurtzite ZnS

Radenka Krsmanović Whiffen [1,2], Amelia Montone [1,*], Loris Pietrelli [3] and Luciano Pilloni [1]

1. ENEA, Materials Technology Division, Casaccia Research Centre, Via Anguillarese 301, 00123 Rome, Italy; radenka.krsmanovic.whiffen@udg.edu.me (R.K.W.); luciano.pilloni@enea.it (L.P.)
2. Faculty of Polytechnics, University of Donja Gorica, Oktoih 1, 81000 Podgorica, Montenegro
3. Department of Chemistry, Sapienza University of Rome, Piazzale Aldo Moro 5, 00185 Rome, Italy; loris.pietrelli@uniroma1.it
* Correspondence: amelia.montone@enea.it

Abstract: Pyroelectric materials can harvest energy from naturally occurring ambient temperature changes, as well as from artificial temperature changes, notably from industrial activity. Wurtzite-based materials have the advantage of being cheap, non-toxic, and offering excellent opto-electrical properties. Due to their non-centrosymmetric nature, all wurtzite crystals have both piezoelectric and pyroelectric properties. Nanocrystalline wurtzite ZnS, being a room temperature stable material, by contrast to its bulk counterpart, is interesting due to its still not well-explored potential in piezoelectric and pyroelectric energy harvesting. An easy synthesis method—a co-precipitation technique—was selected and successfully tailored for nanocrystalline wurtzite ZnS production. ZnS nanopowder with nanoparticles of 3 to 5 nm in size was synthesized in ethyl glycol under medium temperature conditions using $ZnCl_2$ and thiourea as the sources of Zn and S, respectively. The purified and dried ZnS nanopowder was characterized by conventional methods (XRD, SEM, TEM, TG and FTIR). Finally, a constructed in-house pilot plant that is able to produce substantial amounts of wurtzite ZnS nanopowder in an environmentally friendly and cost-effective way is introduced and described.

Keywords: zinc sulfide; wurtzite; co-precipitation synthesis; solvent recycling; green synthesis; scaling up; pilot plant

1. Introduction

The access to and use of electrical energy contributes significantly to sustainable development in modern societies. One possible way to generate green energy is by harvesting waste heat and converting it into electrical energy, using either thermoelectric or pyroelectric effects. While commercial thermo-electric generators exist today [1], "pyroelectric energy harvesting" is the least well-explored potential energy harvesting technology. Pyroelectric materials operate with a high thermodynamic efficiency and do not require bulky heat sinks as thermoelectric materials do. Some pyroelectric materials are also stable at up to 1200 °C or more, permitting energy harvesting from high temperature sources with increased thermodynamic efficiency. As a result, pyroelectric materials have the potential to harvest energy from naturally occurring ambient temperature changes, and artificial temperature changes due to exhaust gases, and gas or liquid hot flows in industrial processes. Thus "pyroelectric energy harvesting" could be the right methodology to collect at least some of the enormous amount of wasted energy in the form of heat by converting thermal fluctuations into electrical energy (e.g., over half the energy generated from all sources in the U.S. since 2008 was reported lost that way [2]).

Pyroelectric materials are structurally anisotropic solids that exhibit a permanent dipole moment, which is why they can generate electricity from temperature fluctuations. As such, they are promising candidates for energy harvesting at a large scale. Conventional materials of this type are ferroelectrics, mostly oxides with a perovskite structure, like

Citation: Krsmanović Whiffen, R.; Montone, A.; Pietrelli, L.; Pilloni, L. On Tailoring Co-Precipitation Synthesis to Maximize Production Yield of Nanocrystalline Wurtzite ZnS. *Nanomaterials* **2021**, *11*, 715. https://doi.org/10.3390/nano11030715

Academic Editor: Sotirios Baskoutas

Received: 8 February 2021
Accepted: 10 March 2021
Published: 12 March 2021

Publisher's Note: MDPI stays neutral with regard to jurisdictional claims in published maps and institutional affiliations.

Copyright: © 2021 by the authors. Licensee MDPI, Basel, Switzerland. This article is an open access article distributed under the terms and conditions of the Creative Commons Attribution (CC BY) license (https://creativecommons.org/licenses/by/4.0/).

BaTiO3 (BT), PbTiO3 (PT), LiTaO$_3$, Lead Magnesium Niobate/Lead Titanate (PMN-PT) or Barium Strontium Titanate (BST) in the forms of single crystals, ceramics or as fillers in polyvinylidene fluoride composite films. Another promising class are non-ferroelectric pyroelectrics: semiconductor materials of wurtzite crystalline structure like CdS, ZnO or ZnS. Due to their non-centrosymmetric nature, all wurtzite crystals have both piezoelectric and pyroelectric properties. Wurtzite-based materials have the advantage of being cheap, non-toxic and offering excellent opto-electrical properties. Their high chemical and thermal stability allow their use at high temperatures in air, whereas ferroelectrics become ineffective when heated beyond their Curie temperature (their T_C is usually lower than 150 °C and can increase at nano-scale [3]). In addition, higher thermal conductivity allows wurtzite-based materials to react faster to ambient temperature change. Currently, the application of pyroelectric materials is limited to low-power electronics, to portable systems or tasks needing only µW–mW power [4]; these applications fit well with nanostructured pyroelectric generators that harvest ambient temperature changes and rapidly generate an electrical current in response to those changes [4]. Although pyroelectric materials and thermal energy harvesting have been studied extensively over the last two decades, research on pyroelectric generators is still in the explorative phase. However, some existing systems such as one developed by NASA [5], are able to successfully capture and transform waste heat generated in power plants, jet engines or automobiles. For more powerful, commercially viable generators that are able to capture industry-generated heat, a more solid form is needed, such as ceramics or thin films with improved mechanical strength and greater resistance to thermal shock.

Alongside ZnO, ZnS is one of the most important Cd-free type II-VI semiconductors with excellent optoelectronic and luminescent properties and numerous applications; in particular, nanostructured ZnS is used mainly as a phosphor in optoelectronic and electro-luminescent devices [6–8], in catalysis, for solar cells, and lasers [9–11] and, being biologically non-toxic, in biomedical labeling [12]. ZnS has two structural forms: a room temperature stable cubic phase (zinc blende or sphalerite, c-ZnS), which at high temperatures (1020 °C for bulk) becomes a metastable, hexagonal wurtzite phase (w-ZnS), lacking structural stability and of limited application. However, a stabilization mechanism for wurtzite is well known: it was found that w-ZnS in nanocrystalline form is a stable material at room temperature [13–16]. These two phases of ZnS have different valence band structure that manifests as a difference in the bandgap value, with widely accepted (experimental) values of 3.72 eV for sphalerite and 3.77 eV for wurtzite [17].

To the best of our knowledge, w-ZnS has not been studied as a possible energy harvesting pyroelectric material since the early work of Gérard Marchal on w-ZnS thin films [18] despite w-ZnS being isostructural to the well-exploited and widely praised hexagonal ZnO [19]. Reducing production costs for pyroelectric material is the key to marketing an affordable energy harvesting system. In addition, the Tc temperature (1020 °C for bulk material) is high enough for ZnS that it has the ability to operate at higher temperature that is a good match with the working temperature of power plants and automobiles (mostly lower than 200 °C, up to maximum about 400–500 °C). We intend to create ceramics and composite polymer films to be used in pyroelectric harvesters of waste heat coming from either industrial or domestic activities [20].

Only in the last decade has the interest in nanostructured wurtzite ZnS sparked, and numerous different syntheses have been explored, aiming at its controlled production at low temperatures [21]. Cheng et al. [22] prepared w-ZnS nanoparticles by a solvothermal method from homogeneous solutions of Zinc chloride ($ZnCl_2$) with S2- as the precipitating anion from thiourea at 180 °C and without a stabilizing agent, while Zhao et al. [23] prepared hexagonal ZnS NPs using $ZnCl_2$ and thiourea controlled by tetramethyl ammonium hydroxide in ethyl glycol (EG). A short list of the possible synthesis is reported in Table S1.

We wanted to avoid the use of toxic materials (Tetramethylammonium hydroxide (TMAH), ethylenediamine, and so on) and to use an inexpensive and simple chemical synthesis at a low working temperature. Hence, we investigated the options of creating

ZnS nanopowder using a co-precipitation fabrication process, a soft-chemistry approach that is easy to scale-up. The obtained nanopowder has been characterized using TGA, BET, FTIR, TEM and SEM techniques. We provided an insight into how to optimize this synthetic route, given conditions for better purity w-ZnS nanopowder, and provided an outlook on how to expand its production. We built an in-house pilot plant that is able to produce substantial amounts of wurtzite ZnS nanopowder in an environmentally friendly and cost-effective way.

2. Materials and Methods

ZnS nanopowder was fabricated using an easy scalable chemical precipitation process. Zn^{2+} ions form a complex with Ethylene glycol (EG) resulting in particle capping upon nucleation. Upon the addition of thiourea (TU) into the preformed Zn-EG complex, a competition between TU and EG (at a high temperature) is performed. We modified the well-known reaction of zinc chloride (Zn^{2+} source) with thiourea (S^{2-} source) dissolved in ethyl glycol (EG) [23] at different molar ratios (R = 0.47–1.22) in medium temperature conditions (140–150 °C) to produce nanocrystalline ZnS of the hexagonal (wurtzite) phase in a series of consecutive experiments. All the chemicals: $ZnCl_2$ (Sigma-Aldrich, St. Louis, MO, USA, $\geq$98%), thiourea (Sigma-Aldrich, St. Louis, MO, USA $\geq$99.0%), and ethylene glycol (Sigma-Aldrich, St. Louis, MO, USA $\geq$99.8%), were used as received and without further purification.

Typically, a known quantity of anhydrous $ZnCl_2$ and thiourea (CH_4N_2S) were dissolved separately in a known volume of EG (the ratio solid:liquid = 1:15) and stirred at 110 °C for 30 min. Subsequently, both mixtures were merged in a larger glass beaker and stirred (300–400 rpm) at a temperature of 140–150 °C for 1–2 h. After the reaction was complete, the white solution was cooled to room temperature. The powders obtained were separated by centrifugation, at 4000 rpm for 5 min, from the solvent and were washed twice with acetone and twice with ethyl alcohol using centrifugation. The washed product was mixed in ethyl alcohol and this solution was dried to powder at 70 °C, in an oven for about 1 h. Following the lab tests, we built an in-house pilot plant able to produce substantial amounts of wurtzite ZnS nanopowder in an environmentally friendly and cost-effective way. To be specific, the pilot plant consists of a 5 L jacketed glass reactor equipped with the automatic control of pH, temperature and mixing speed. The temperature was controlled ($\pm$0.01 °C) for the working temperature range 10–200 °C by a thermostat equipped with a programmable temperature fluid control (model Optima TXF200 Heated Circulating Bath).

The chemical process follows the scheme reported in Figure 1.

The microstructure of the ZnS nanopowders was characterized by X-ray powder diffraction using a SmartLab Rigaku powder diffractometer (Rigaku, Tokio, Japan) equipped with a Cu Kα radiation source and a graphite monochromator in the diffracted beam, operated at 40 kV and 30 mA. The morphology of the samples was investigated by scanning electron microscopy, using a LEO 1530 (Zeiss, Oberkochen, Germany)) instrument. The LEO1530 is a hot cathode field emission SEM equipped with a high-resolution in-lens secondary electron detector, a conventional secondary electron detector, a Centaurus back scattered detector and a XACT microanalysis unit (OXFORD Instruments, Abingdon, United Kingdom) and it was used to provide high-resolution images. TEM images were obtained with a JEOL 2010 TEM (Jeol, Akishima, Japan).

The TG measurements were carried out by a Mettler Toledo thermogravimetric analyzer (Mettler Toledo, Columbus, OH, USA) under a nitrogen atmosphere, where the gas flow was fixed at 20 mL min^{-1}. The heating rate was fixed at 5.0 °C min^{-1} and the samples (3–6 mg) were placed in an alumina crucible.

The UV absorption spectra were taken using a Shimadzu UV-1800 Spectrophotometer (Shimadzu, Kyoto, Japan) in the wavelength range 200–850 nm.

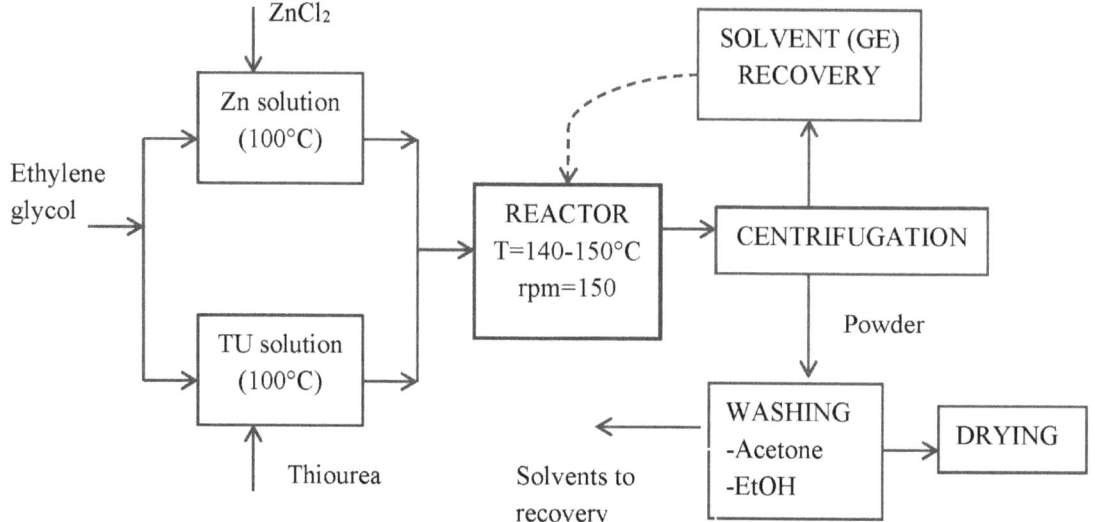

Figure 1. The schematic representation of the ZnS nanopowder synthesis.

The measurement of the specific surface area of the produced wurtzite ZnS powder was carried out using the Brunauer-Emmett-Teller (BET) equation. The samples (about 1 g) were dried in a vacuum system at 120 °C overnight. Nitrogen was used as an adsorbate gas at 77 K (Nova 2200 surface-area analyzer; Quantachrome Instruments, Boynton Beach, FL, USA).

The infrared transmittance measurements of the produced powder were performed using a Thermo Scientific Nicolett 6700 spectrophotometer (Thermo Fisher Scientific, Waltham, MA, USA) at room temperature; the spectrum range was 4000–400 cm^{-1} wavenumber range and a resolution of 2 cm^{-1}.

The elemental analysis was performed in order to verify the purity of the powder using a Carlo Erba Instruments EA 1110 CHNS-O elemental analyzer (Egelsbach, Germany).

3. Results and Discussion

3.1. On Synthesis

In order to produce wurtzite ZnS in an environmentally friendly and cost-effective way, we slightly modified the reaction by employing simple mixing at atmospheric pressure instead of a solvothermal method and, moreover, by recycling the solvent that still contained Zn^{2+} ions. The w-ZnS nanoparticles were synthesized by a co-precipitation reaction using precursor solutions at different Zn^{2+}/S^{2-} ratios.

In order to minimize waste according to the circular economy approach in the preparation of high value compounds such as w-ZnS NPs, following the recommendations of the European Environmental Agency [24], the synthesis was designed for reuse with both the reaction media (GE) and the washing solvent.

We observed different color formations in the reactive solution during the synthesis process. The starting solution was transparent, and was heated up gradually. At 140 °C, a milky white solution was developed that transformed to rose pink almost instantaneously once the temperature reached 150 °C, as shown in Figure 2.

Figure 2. The color of the solution changes with temperature from milky white at 140 °C to rose pink at 150 °C. This change was very quick and easily detectable.

The pink color observed at about 150 °C is probably due to the formation of a zinc complex favored by the isomerization of thiourea in ammonium thiocyanate. A reversible reaction (1) of thiourea isomerization into thiocyanate (NH_4SCN) occurring in the range 140–180 °C with an equilibrium ratio of TU:NH_4SCN at 1:3 [25]. With the increasing temperature (>150 °C), the formation of guanidinium thyocyanate can occur according to reaction (3):

$$SC(NH_2)_2 \leftrightarrow NH_4SCN \qquad (1)$$

$$SC(NH_2)_2 \rightarrow NH_2CN + H_2S \qquad (2)$$

$$NH_4SCN + NH_2CN \rightarrow [H_2N^+=C\,(NH_2)_2]\,SCN^- \qquad (3)$$

The ammonium thiocyanate Formation (1) can promote the precipitation of the tetrathiocyanatozincate(II) anion complex [26].

The experimental results of TU decomposition by TGA indicate that NH_4SCN, H_2S, NH_3, CS_2, HNCS can be formed [27,28].

Figure 3 presents the TGA curves of the produced w-ZnS; the curves show mass loss in the temperature range 200–330 °C probably due to the production of the carbon disulfide (CS_2) and ammonia (NH_3), while at T > 500 °C, we expect other gaseous species such as cyanamide (H_2NCN), hydrogen cyanide (HCN), carbon dioxide (CO_2) and carbonyl sulfide (COS) to be formed, as it was reported by Madaraz and Pokol et al. [28].

The complete decomposition of the thio- and cyano- compounds occurs at T > 500 °C, where a weight loss of between 15% and 31% was observed (see Figure 3).

The FTIR analysis of w-ZnS powders is shown in Figure 4. The spectrum obtained from the slightly pink colored ZnS powder synthesized at 150 °C show at 1100–1500 nm, the characteristic absorption bands of the sulphur compounds such as C=S, CSNH, SO_2, SO_2N, while at 2060 nm, the SCN vibrations and, at 1400 nm, the NH_4 peak are present. These data confirm the presence of TU degradation compounds as observed by TGA. More research will be conducted to study this degradation mechanism in greater detail [25]. The broad absorption band at 3200 nm can be attributed to OH from the adsorbed H_2O on the surface of the powder, while the bands at 1560 nm and 1428 nm can be attributed to the zinc carboxylate [29].

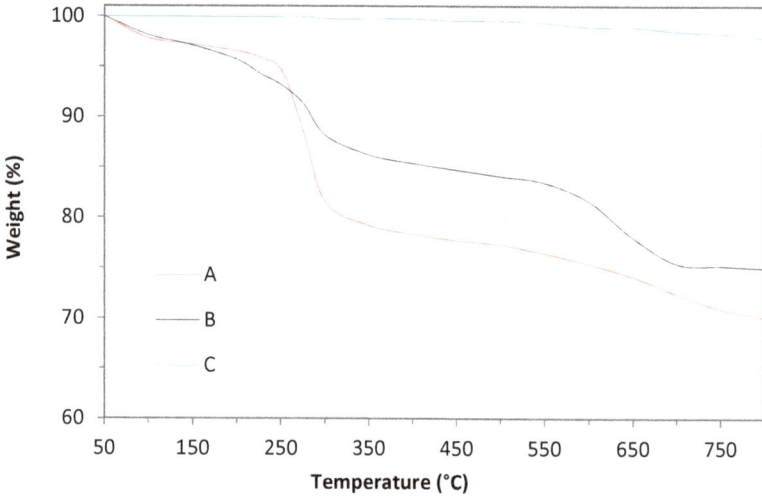

Figure 3. The TGA curves of the thermal decomposition of the produced wurtzite ZnS. Red (A) = washed with water for 5 min, black (B) = washed with acetone and ethanol for 5 min, blue (C) = commercial ZnS powder (UMICORE). The flow rate of N_2 was 20 mL min^{-1}; the heating rate equals 5 °C min^{-1}.

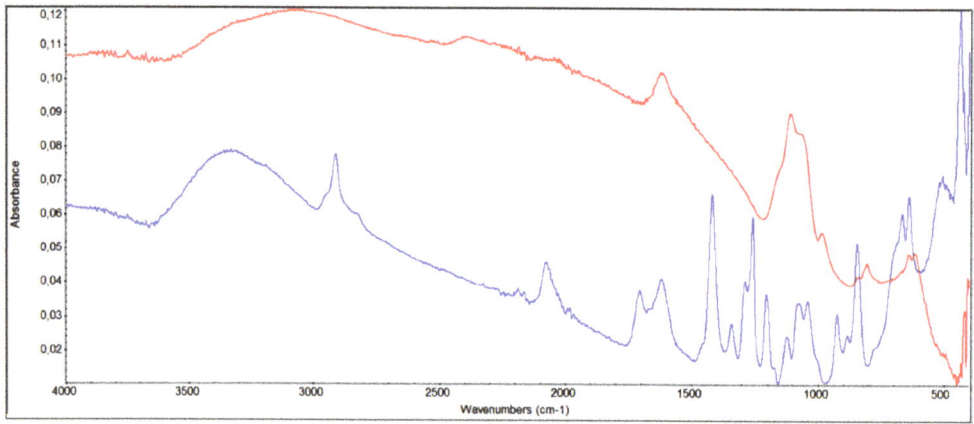

Figure 4. The FTIR spectra of w-ZnS washed solution (acetone and ethanol) for different time: 5 min (blue) and 30 min (red).

On the other hand, the comparison of the UV spectra of the washing solutions (wurtzite nanopowder in washing medium acetone and ethyl alcohol) used for the ZnS nanopowders obtained at 140 °C and 150 °C (Figure S1 in Supplementary Material) showed the presence of a peak (around 320 nm) in the spectrum of the solvent used for the powder obtained at 150 °C, probably attributable to a derivative of the carbodiimide which is one of the products of the decomposition of thiourea.

The elemental analysis shows that after washing the resultant w-ZnS nanopowder twice with acetone and twice with EtOH, impurities are still present. In particular, the powder contains C = 13.77%, N = 1.18% and H = 2.50%. By introducing an additional washing step with cold water, the following elemental analysis was obtained: C = 7.11%, N = 0.34% and H = 1.55%. A successful removal of the solvents and degradation products was achieved by increasing the washing times, as reported in Figure 4.

The effect of the nmZn/nMS molar ratio can be observed in Figure S2 of the Supplementary Material. The quantity of the nanopowder production is strongly correlated to the concentration of zinc ions; in particular, an excess of zinc ions is required to increase the production of w-ZnS nanopowder.

3.2. Temperature Effect

Temperature plays an important role in the synthesis of nano w-ZnS. According to Cheng et al. [19], the Zn^{2+} ions and TU homogeneous EG solutions are mixed, and the following reaction can be described:

$$Zn^{2+} + nCS(NH_2)_2 \rightarrow \{Zn[CS(NH_2)_2]_n\}^{2+} \quad (4)$$

creating a strong coordination complex that at about 110 °C may decompose generating nucleuses and then ZnS crystals having hexagonal form.

Moreover, to avoid obtaining a precipitate containing degradation products of thiourea that are very difficult to wash out, it is necessary that the synthesis reaction take place at 140 °C. We performed dozens of experiments and we can confirm that synthesis at 140 °C is the only way to obtain a white w-ZnS nanopowder as the final product which contains only the solvent (EG) and the reagents (TU and $ZnCl_2$) as impurities, which are easy to remove through a standard washing procedure using centrifugation.

We also investigated options involving tailoring this synthesis route to maximize the production yield of nanocrystalline wurtzite ZnS. We used the same reaction of zinc chloride with thiourea dissolved in ethyl glycol to produce pure, nanocrystalline ZnS of hexagonal phase in a series of consecutive experiments. The amount of the solvent was kept the same (60 mL of ethyl glycol) by re-using the remnants of the solvent from the previous reaction and topping up the quantity lost. The productivity yield increased 3.5 times in 6 successive reactions, from 156 mg to 549 mg per batch at a constant ratio R = mMZn/mMS = 1 (see Figure 5c). The solvent in the last batch solution contained 17 ppm of zinc and 6 ppm of sulphur. From the XRD measurements in Figure S3, we can see that the "standard" sample is more crystalline in respect to the "recycled" ones, as the latter contain more remnants from the organic part, as confirmed by the TGA (Figure 3) and FTIR (Figure 4) measurements.

It is also worth noting that the pilot plant is able to create a considerable amount of nanopowder relatively quickly: across 3 batches produced using the same solvent, (V_{EG} = 600 mL, t = 2 h, T = 140 °C, molar ratio mMZn/mMS $\approx$ 0.45) 13.732, 11.437, and 12.685 g of wurtzite ZnS were obtained, giving a total of 37.854 g for the whole process.

3.3. Structural and Microstructural Characterization

With the synthesis method described above, we were able to produce pure, nanocrystalline ZnS of hexagonal (wurtzite) phase (Figure 5b) in a series of consecutive experiments. After the final drying procedure of the ZnS solution, we observed an unusual phenomenon: a self-alignment of the ZnS wurtzite nanopowder in highly ordered arrays (Figure 5a), probably arising due to the inherent polar nature of the wurtzite nanoparticles and the polar nature of the solvent (ethyl alcohol).

XRD analysis confirmed the hexagonal structure of ZnS (see Figure 5d), while SEM observation (Figure 6) showed nicely agglomerated spheres made of nanoparticles that we could clearly observe in the TEM image (Figure 7). The SEM observations revealed that the nanopowder sample is organized quite uniformly on a large scale in 100–200 nm-size globular structures (see Figure 6). The microstructure at the local level was studied using TEM. HRTEM images taken at higher magnification show that the w-ZnS samples are nanophase materials with crystallite made up of about 3 nm in size. The specific surface area of the ZnS nanopowder was measured as being 38 m^2/g.

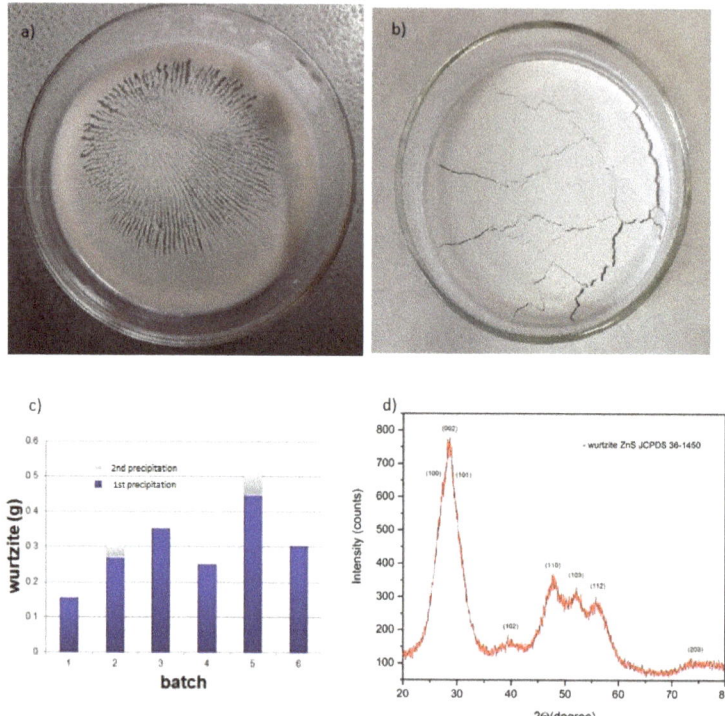

Figure 5. (**a**) Self-alignment of the ZnS wurtzite nanopowder (after thermal treatment at 70 °C for 1 h) into highly ordered arrays; (**b**) The final product—wurtzite ZnS nanopowder; (**c**) The results of the "recycling" synthesis experiments; and (**d**) The XRD diffractogram taken from the produced nanopowder ZnS.

Figure 6. SEM image. The obtained ZnS is constituted by agglomerates composed by nanoparticles whose dimensions are in the range of a few nanometers.

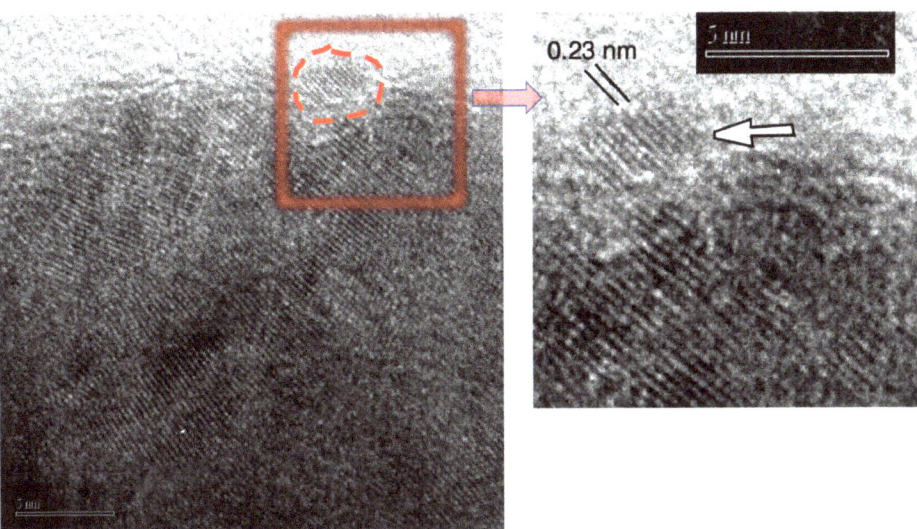

Figure 7. High-resolution TEM image. The images show nanoparticles as small as about 3 nm taken near the surface of an agglomerate. The measured distance between plains is compatible with the distance of (1,0,2) planes of hexagonal ZnS (i.e., about 0.23 nm).

4. Conclusions

In order to produce wurtzite in an environmentally friendly and cost-effective way, a co-precipitation reaction including a simple mixing of precursors at atmospheric pressure was explored. In addition, a successful procedure for recycling of the solvent that still contains Zn^{2+} ions (from the previous reaction) was introduced. As a consequence of this greener, "circular approach", that complements ongoing research trends towards eco-friendly nanoparticle production [30], the productivity yield increased 3.5 times. Following the lab tests, an in-house pilot plant was built, able to produce substantial amounts of wurtzite ZnS nanopowder in an environmentally friendly and cost-effective way, whereby, across three batches prepared in sequence using the same solvent, approximately 38 g were obtained (V_{EG} = 600 mL, t = 2 h, T = 140 °C, mMZn/mMS ≈ 0.45). The yield increased using this molar ratio as reported in Figure S2. Based on the present results, the main advantages of our home-made pilot plant (see Figure S4 of Supplementary Material) include: (i) easy assemblage from commercially available items, (ii) transparency of the glass reactor for clear monitoring of the synthesis process, (iii) production yield of about 18–20% per batch, and (iv) 100% solvent recyclability.

A further investigation regarding the washing procedure of the synthetized w-ZnS nanopowder is envisaged to achieve a higher degree of purity, for example, the use of an ultrasound bath could lead to a greater removal of impurities while reducing the washing time. Our thermogravimetric measurements show that the remnants of ethyl glycol and thiourea decompose completely at 250–290 °C (see Figure 3), hence a flash-heating of the powder might be worth exploring as an additional purification method.

One crucial aspect that highlights the importance of recycling within the chemical process is the solvent recovery and reuse. Our practice allows for a significant reduction in the amount of solvent needed for the w-ZnS nanopowder production process, and we expect to employ the same approach for the synthesis of other similar materials such as ZnO.

Supplementary Materials: The following are available online at https://www.mdpi.com/2079-4991/11/3/715/s1, Table S1: Synthesis of wurtzite ZnS by co-precipitation technique, Figure S1: The UV absorption spectra of the w-ZnS powder washing solutions: w-ZnS produced at 150 °C (blue), at 140 °C (red) and of clean solvent (green), Figure S2: The graph showing the production of w-ZnS (in grams) as a function of the nmZn/nMS molar ratio used in the synthesis, Figure S3: XRD diffractogram taken from the ZnS "standard" samples—red and blue, and from the "recycled" samples—green, orange and purple lines, Figure S4: The glass reactor of the pilot plant and jars containing the recycled solvent (right) and the w-ZnS solution (left). The pilot plant consists of a 5 L transparent jacketed glass reactor with the mechanical stirrer, equipped with a temperature sensor and controller, stirring velocity controller and a pH value indicator, as well as a circulating bath with advanced digital temperature controller.

Author Contributions: Conceptualization, R.K.W. and L.P. (Loris Pietrelli); methodology, L.P. (Loris Pietrelli); validation, R.K.W., L.P. (Loris Pietrelli) and A.M.; formal analysis, R.K.W. and L.P. (Loris Pietrelli); investigation, R.K.W., L.P. (Loris Pietrelli) and L.P. (Luciano Pilloni); data curation, R.K.W., A.M. and L.P. (Loris Pietrelli); writing—original draft preparation, R.K.W.; writing—review and editing, L.P. (Loris Pietrelli), L.P. (Luciano Pilloni) and A.M.; visualization, R.K.W. and L.P. (Loris Pietrelli); resources, supervision, project administration, A.M.; funding acquisition, R.K.W. and A.M. All authors have read and agreed to the published version of the manuscript.

Funding: This project has received funding from the European Union's Horizon 2020 research and innovation programme under the Marie Skłodowska-Curie grant agreement number 797951. This project was also partially supported by the Piano triennale di realizzazione 2019–2021 della ricerca di sistema elettrico nazionale—Progetto 1.3 Materiali di frontiera per usi energetici (C.U.P. code: I34I19005780001).

Data Availability Statement: Data is contained within the article or supplementary material.

Conflicts of Interest: The authors declare no conflict of interest.

References

1. Thermoelectric Generator Power Products for Sale. Available online: https://tecteg.com/thermoelectric-generator-power-products-for-sale/ (accessed on 15 January 2021).
2. Lawrence Livermore National Laboratory "Estimated Energy Use in 2015: 97.5 Quads". Available online: https://flowcharts.llnl.gov (accessed on 20 December 2020).
3. Li, Y.; Liao, Z.; Fang, F.; Wang, X.; Li, L.; Zhu, J. "Significant increase of Curie temperature in nano-scale BaTiO3". *Appl. Phys. Lett.* **2014**, *105*, 182901. [CrossRef]
4. Morozovska, A.N.; Eliseev, E.A.; Svechnikov, S.; Kalinin, V. Pyroelectric response of ferroelectric nanowires: Size effect and electric energy harvesting. *J. Appl. Phys.* **2010**, *108*, 042009. [CrossRef]
5. Pyroelectric Sandwich Thermal Energy Harvesters US10147863B2 (A1) • 2018-12-04 • NASA [US], Earliest Priority: 2014-10-09 • Earliest Publication: 2016-04-14. Available online: http://patft.uspto.gov/netacgi/nph-Parser?Sect1=PTO1&Sect2=HITOFF&d=PALL&p=1&u=%2Fnetahtml%2FPTO%2Fsrchnum.htm&r=1&f=G&l=50&s1=10,147,863.PN.&OS=PN/10,147,863&RS=PN/10,147,863 (accessed on 15 November 2020).
6. Shionoya, S.; Yen, W.M. *Phosphor Handbook*; CRC Press LLC: Boca Raton, FL, USA, 1999.
7. Sarkar, R.; Tiwary, C.; Kumbhakar, P.; Basu, S.; Mitra, A. Yellow-orange light emission from Mn2+-doped ZnS nanoparticles. *Phys. E: Low-dimensional Syst. Nanostructures* **2008**, *40*, 3115–3120. [CrossRef]
8. Whiffen, R.K.; Jovanović, D.; Antić, Ž.; Bártová, B.; Milivojević, D.; Dramićanin, M.; Brik, M. Structural, optical and crystal field analyses of undoped and Mn2+-doped ZnS nanoparticles synthesized via reverse micelle route. *J. Lumin* **2014**, *146*, 133–140. [CrossRef]
9. Fang, X.; Bando, Y.; Gautam, U.K.; Zhai, T.; Zeng, H.; Xu, X.; Liao, M.; Golberg, D. ZnO and ZnS Nanostructures: Ultraviolet-Light Emitters, Lasers, and Sensors. *Crit. Rev. Solid State Mater. Sci.* **2009**, *34*, 190–223. [CrossRef]
10. Wang, X.; Huang, H.; Liang, B.; Liu, Z.; Chen, D.; Shen, G. ZnS Nanostructures: Synthesis, Properties, and Applications. *Crit. Rev. Solid State Mater. Sci.* **2013**, *38*, 57–90. [CrossRef]
11. Tiwari, A.; Dhoble, S.J. Critical Analysis of Phase Evolution, Morphological Control, Growth Mechanism and Photophysical Applications of ZnS Nanostructures (Zero-Dimensional to Three-Dimensional): A Review. *Cryst. Growth Des.* **2017**, *17*, 381–407. [CrossRef]
12. Zhang, F.; Li, C.; Li, X.; Wang, X.; Wan, Q.; Xian, Y.; Jin, L.; Yamamoto, K. ZnS quantum dots derived a reagentless uric acid biosensor. *Talanta* **2006**, *68*, 1353–1358. [CrossRef] [PubMed]
13. Wang, Z.; Daemen, L.L.; Zhao, Y.; Zha, C.S.; Downs, R.T.; Wang, X.; Wang, Z.L.; Hemley, R.J. Morphology-tuned wurtzite-type ZnS nanobelts. *Nat. Mater.* **2005**, *4*, 922–927. [CrossRef]

14. Zhao, Z.; Geng, F.; Cong, H.; Bai, J.; Cheng, H.-M. A simple solution route to controlled synthesis of ZnS submicrospheres, nanosheets and nanorods. *Nanotechnology* **2006**, *17*, 4731–4735. [CrossRef] [PubMed]
15. Fang, X.; Zhai, T.; Gautam, U.K.; Li, L.; Wu, L.; Bando, Y.; Golberg, D. ZnS nanostructures: From synthesis to applications. *Prog. Mater. Sci.* **2011**, *56*, 175–287. [CrossRef]
16. Huo, F.; Wang, Y.; You, C.; Deng, W.; Yang, F.; Pu, Y. Phase- and size-controllable synthesis with efficient photocatalytic activity of ZnS nanoparticles. *J. Mater. Sci.* **2017**, *52*, 5626–5633. [CrossRef]
17. Dong, M.; Zhang, J.; Yu, J. Effect of effective mass and spontaneous polarization on photocatalytic activity of wurtzite and zinc-blende ZnS. *APL Mater.* **2015**, *3*, 104404. [CrossRef]
18. Marchal, G. Pyroélectricité du sulfure de zinc en couches minces. *J. Phys.* **1970**, *31*, 779–782. [CrossRef]
19. Yang, Y.; Guo, W.; Pradel, K.C.; Zhu, G.; Zhou, Y.; Zhang, Y.; Hu, Y.; Lin, L.; Wang, Z.L. Pyroelectric Nanogenerators for Harvesting Thermoelectric Energy. *Nano Lett.* **2012**, *12*, 2833–2838. [CrossRef]
20. Chavez, L.A.; Jimenez, F.O.Z.; Wilburn, B.R.; Delfin, L.C.; Kim, N.; Love, N.; Lin, Y. Characterization of Thermal Energy Harvesting Using Pyroelectric Ceramics at Elevated Temperatures. *Energy Harvest. Syst.* **2018**, *5*, 3–10. [CrossRef]
21. La Porta, F.A.; Andrés, J.; Li, M.S.; Sambrano, J.R.; Varela, J.A.; Longo, E. Zinc blende versus wurtzite ZnS nanoparticles: Control of the phase and optical properties by tetrabutylammonium hydroxide. *Phys. Chem. Chem. Phys.* **2014**, *16*, 20127–20137. [CrossRef]
22. Cheng, Y.; Lin, Z.; Lu, H.; Zhang, L.; Yang, B. ZnS nanoparticles well dispersed in ethylene glycol: Coordination control synthesis and application as nanocomposite optical coatings. *Nanotechnology* **2014**, *25*, 115601. [CrossRef]
23. Zhao, Y.; Zhang, Y.; Zhu, H.; Hadjipanayis, A.G.C.; Xiao, J.Q. Low-Temperature Synthesis of Hexagonal (Wurtzite) ZnS Nanocrystals. *J. Am. Chem. Soc.* **2004**, *126*, 6874–6875. [CrossRef]
24. European Environment Agency. *Circular by Design Products in the Circular Economy*; Publications Office of the European Union: Luxembourg, 2017.
25. Timchenko, V.P.; Novozhilov, A.L.; Slepysheva, O.A. Kinetics of Thermal Decomposition of Thiourea. *Russ. J. Gen. Chem.* **2004**, *74*, 1046–1050. [CrossRef]
26. Rao, T.P.; Ramakrishna, T.V. Spectrophotometric determination of zinc with thiocyanate and Rhodamine 6G. *Analyst* **1980**, *105*, 674–678. [CrossRef]
27. Wang, S.; Gao, Q.; Wang, J. Thermodynamic Analysis of Decomposition of Thiourea and Thiourea Oxides. *J. Phys. Chem. B* **2005**, *109*, 17281–17289. [CrossRef]
28. Madarász, J.; Pokol, G. Comparative evolved gas analyses on thermal degradation of thiourea by coupled TG-FTIR and TG/DTA-MS instruments. *J. Therm. Anal. Calorim.* **2007**, *88*, 329–336. [CrossRef]
29. Labiadh, H.; Lahbib, K.; Hidouri, S.; Touil, S.; Ben Chaabane, T. Insight of ZnS nanoparticles contribution in different biological uses. *Asian Pac. J. Trop. Med.* **2016**, *9*, 757–762. [CrossRef]
30. Reverberi, A.; Vocciante, M.; Lunghi, E.; Pietrelli, L.; Fabiano, B. New Trends in the Synthesis of Nanoparticles by Green Methods. *Chem. Eng. Trans.* **2017**, *61*, 667–672. [CrossRef]

Article

CO₂ Hydrogenation over Unsupported Fe-Co Nanoalloy Catalysts

Marco Calizzi [1,2], Robin Mutschler [1,2], Nicola Patelli [3,*], Andrea Migliori [4], Kun Zhao [1,2], Luca Pasquini [3] and Andreas Züttel [1,2]

1. Laboratory of Materials for Renewable Energy, Institute of Chemical Sciences and Engineering, École Polytechnique Fédérale de Lausanne, 1951 Sion, Switzerland; marco.calizzi@gmail.com (M.C.); robin.mutschler@empa.ch (R.M.); kun.zhao@epfl.ch (K.Z.); andreas.zuettel@epfl.ch (A.Z.)
2. EMPA Materials Science & Technology, 8600 Dübendorf, Switzerland
3. Department of Physics and Astronomy, Alma Mater Studiorum Università di Bologna, 40127 Bologna, Italy; luca.pasquini@unibo.it
4. Unit of Bologna, Institute of Microelectronics and Microsystems, National Research Council, 40129 Bologna, Italy; migliori@bo.imm.cnr.it
* Correspondence: nicola.patelli@unibo.it

Received: 19 June 2020; Accepted: 9 July 2020; Published: 11 July 2020

Abstract: The thermo-catalytic synthesis of hydrocarbons from CO_2 and H_2 is of great interest for the conversion of CO_2 into valuable chemicals and fuels. In this work, we aim to contribute to the fundamental understanding of the effect of alloying on the reaction yield and selectivity to a specific product. For this purpose, Fe-Co alloy nanoparticles (nanoalloys) with 30, 50 and 76 wt% Co content are synthesized via the Inert Gas Condensation method. The nanoalloys show a uniform composition and a size distribution between 10 and 25 nm, determined by means of X-ray diffraction and electron microscopy. The catalytic activity for CO_2 hydrogenation is investigated in a plug flow reactor coupled with a mass spectrometer, carrying out the reaction as a function of temperature (393–823 K) at ambient pressure. The Fe-Co nanoalloys prove to be more active and more selective to CO than elemental Fe and Co nanoparticles prepared by the same method. Furthermore, the Fe-Co nanoalloys catalyze the formation of C_2-C_5 hydrocarbon products, while Co and Fe nanoparticles yield only CH_4 and CO, respectively. We explain this synergistic effect by the simultaneous variation in CO_2 binding energy and decomposition barrier as the Fe/Co ratio in the nanoalloy changes. With increasing Fe content, increased activation temperatures for the formation of CH_4 (from 440 K to 560 K) and C_2-C_5 hydrocarbons (from 460 K to 560 K) are observed.

Keywords: nanoparticle; nanoalloy; catalyst; CO_2 reduction; hydrocarbon; synthetic fuel; iron; cobalt

1. Introduction

CO_2 capture and utilization (CCU) is the process of capturing CO_2 anthropogenic emissions and using them to synthesize valuable and useful chemicals. When applied to fuel production, this concept translates into a closed carbon cycle, thus implementing a sustainable energy system. The production of liquid synthetic fuels is especially interesting for large scale energy storage because they retain all the benefits of liquid fossil fuels, such as high energy density and stability in ambient conditions [1,2]. In this framework, the quest for a material that efficiently catalyzes the reaction between CO_2 and H_2 is of key importance, since CO_2 is a very stable molecule ($\Delta_f H^0_{298\,K/CO_2} = -393.5$ kJ/mol).

Historically, there are two major reactions for the thermo-catalytic synthesis of hydrocarbons from CO_2 or CO: The Sabatier reaction (Equation (1)), which is highly selective towards CH_4, and the Fischer–Tropsch (FT) reaction (Equation (2)), which is up to date and the most industrially relevant reaction to synthesize hydrocarbon fuels, alcohols and waxes from syngas (mixture of H_2 and CO, mainly). If CO_2 obtained from the atmosphere or from local emitters is the starting molecule for the synthesis, the FT reaction can be combined with the (endothermic) reverse water gas-shift reaction (RWGS, Equation (3)). Another variation is the direct conversion of CO_2 to higher hydrocarbons via a FT-like reaction (Equation (4)).

$$CO_2 + 4H_2 \rightarrow CH_4 + 2H_2O \quad (\Delta H^0 = -164.9 \text{ kJ/mol}), \tag{1}$$

$$nCO + (2n+1)H_2 \rightarrow C_nH_{2n+2} + nH_2O, \tag{2}$$

$$CO_2 + H_2 \rightarrow CO + H_2O \quad (\Delta H^0 = +41.18 \text{ kJ/mol}), \tag{3}$$

$$nCO_2 + (3n+1)H_2 \rightarrow C_nH_{2n+2} + 2nH_2O, \tag{4}$$

The C_{2+} products are energetically close together, therefore, the catalyzed synthesis leads to a wide range of products. Furthermore, the synthesis of C_{2+} products requires the reaction of CO_2 with hydrogen and simultaneously also the reaction between the C-containing intermediates. These two competing reactions have to be controlled independently in order to yield a specific product. The direct CO_2 hydrogenation is of great importance and recent publications have shown that the selectivity and yield towards C_{2+} hydrocarbons can be increased either via the combination of different catalysts in a multi-catalysts bed [3–5], or via alloying the transition metals Fe, Co, Ni and Cu, combined with alkali metal promoters, such as Na and K [6].

Fe-based alloys, such as Fe-M (M = Cu, Co, Ni), have been in the focus of investigations since they were found to be the most active elements of industrial relevance in the formation of longer chained HCs via direct CO_2 hydrogenation [7–9]. Among these kinds of materials, supported Fe-Co-based nanoparticles (NPs) show the highest C_{2+} yields: 25.4% C_{2+} yield with 35.8% CO_2 conversion for K-promoted $Fe_{0.9}Co_{0.1}$ on Al_2O_3 [10], or 14.3% C_{2+} yield with 33.3% CO_2 conversion for $Fe_{0.9}Co_{0.1}$ on TiO_2 (1.1 MPa, 573 K) [9].

While the supporting metal oxide phases are known to alter the reaction, they also provide mechanical and thermal stability to the NPs to avoid sintering [11]. The mainly alkali metal promoters are used to enhance the selectivity towards C_{2+} hydrocarbons.

The principal focus of the literature is on finding the material with the best catalytic activity and selectivity in order to increase the reaction yield. In this work, instead, we aim to contribute to the fundamental understanding of the effect of alloying on the reaction yield and selectivity to a specific product. Therefore, to study the fundamental influence of alloying Fe and Co, we relinquish the use of a metal oxide support and promoters. The aim of this paper is the analysis of the structure, composition and stability of unsupported Fe-Co alloy NPs (nanoalloys), synthesized via Inert Gas Condensation (IGC), and their catalytic properties in the CO_2 hydrogenation reaction.

IGC is chosen as synthesis method since it is a versatile technique that can produce elemental NPs [12], bimetallic NPs [13,14], and other hydride or oxide nanocomposites [15,16] of high purity, with good control of the bulk composition in free-standing powder form.

2. Materials and Methods

NPs were grown via IGC in an ultrahigh-vacuum (UHV) chamber, equipped with a tungsten boat as a thermal vapor source [17]. The precursor material of the Fe-Co nanoalloys was a mixture of microcrystalline Fe (Sigma-Aldrich, Darmstadt, Germany, particle size <450 µm, purity >99%) and Co (Sigma-Aldrich, particle size <150 µm, purity ≥99.9%) powders previously melted and alloyed in the tungsten boat. Three Fe-Co powder mixtures with 25 wt%, 50 wt% and 75 wt% Co content were used to synthesize nanoalloys with different Fe/Co ratios. The corresponding samples are named 30Fe70Co, 50Fe50Co, and 76Fe24Co throughout the manuscript, according to the SEM-EDX quantified elemental wt% content of the sample, as reported in Table 1. Single-element Fe and Co NPs were also synthesized from the unmixed powders for comparison.

Table 1. Mean crystallite size $\overline{d}$ and lattice parameter a obtained from XRD analysis of metallic bcc reflections; Fe and Co content from EDX; BET-measured surface area S_A^{BET}; surface area S_A^{TEM} and volume-weighted mean diameter $\overline{d}^{TEM}$ calculated from TEM size distributions. The numbers in parenthesis represent the standard error in units of the last significant digit.

Sample	$\overline{d}$	a (Å)	wt% Fe	wt% Co	S_A^{BET} (m^2 g^{-1})	S_A^{TEM} (m^2 g^{-1})	$\overline{d}^{TEM}$ (nm)
Fe	15 (1)	2.8729 (4)	100	0	–	–	–
76Fe24Co	19 (1)	2.8686 (2)	76 (2)	24 (2)	47 (9)	56	15 (7)
50Fe50Co	18 (1)	2.8629 (2)	50 (2)	50 (2)	56 (5)	63	13 (5)
30Fe70Co	25 (1)	2.8410 (2)	30 (2)	70 (2)	56 (5)	37	22 (7)

The synthesis chamber was preliminarily evacuated to 2×10^{-5} Pa and then filled with He (99.9999% purity) up to a final pressure of 260 Pa. NPs nucleated in the gas phase because the metal vapors quickly supersaturated via thermalization with the surrounding He gas. A He flow of 60 mL$_n$/min, regulated with the aid of a mass flow controller, was directed on the evaporation boat during the whole synthesis in order to facilitate the removal of the NPs from the hot zone where NP coalescence takes place. The pressure was maintained constant by the simultaneous operation of the rotary pump. A liquid N$_2$ cooled rotating stainless-steel cylinders allowed for the collection of NPs via thermophoresis. The NPs were finally scraped off by means of a stainless-steel blade and transferred into an auxiliary UHV chamber, from which they were extracted and sealed under inert Ar atmosphere.

The composition and structure of the as-prepared NPs were determined by scanning electron microscopy (SEM) with a Leica Cambridge Stereoscan 360 equipped with an X-ray detector for energy dispersive X-ray (EDX) microanalysis, and by X-ray diffraction (XRD) with a PANalytical X'celerator diffractometer employing Cu Kα radiation. XRD patterns were analyzed with the MAUD Rietveld refinement software [18] to determine the lattice parameters, crystallite size, and relative phase abundance.

The morphology and the size distribution of the samples were analyzed by means of a FEI Tecnai F20 ST transmission electron microscope (TEM), operated at 200 kV. High-angle annular dark field (HAADF) images and EDX elemental profiles at the single NP level were acquired in the scanning TEM mode (STEM). For TEM analysis, the NPs were dispersed in isopropanol and the suspension was drop-casted on a holey carbon support grid. The specific surface areas of the nanoalloys were determined by applying the Brunauer–Emmett–Teller (BET) method to N$_2$ adsorption isotherms measured in a Belsorp mini II instrument after degassing in a vacuum at 423 K for 0.5 h.

The CO$_2$ hydrogenation experiments were carried out in a dedicated gas control and analysis system with a highly isothermal packed bed [19]. The reactor was loaded with 10 mg of catalyst in the glovebox and the reactor tubing was pumped out and flushed with He three times before the valves that connect the reactor to the tubing were opened. After opening the valves, a constant He flow of 10 mL$_n$/min was applied and the reactor was heated up to 393 K.

The catalyst and the reactor tubing were pre-treated in these conditions for at least 30 min to evaporate remaining moisture. Furthermore, all the tubes in the downstream were heated to 433–473 K in order to evaporate the moisture in the tubing to the mass spectrometer. After the He pretreatment, the catalyst and tubing were treated with a 7.5 mL$_n$/min H$_2$ and 2.5 mL$_n$/min He gas mixture at 393 K to further reduce remaining oxides. The nanoalloys were not pre-reduced at high temperature to avoid coarsening phenomena that would lead to the loss of the nanostructure. They were dispersed on glass-wool to facilitate the loading into the reactor and to distribute evenly in the reactor tube.

The aluminum inlet in the reactor oven ensured a uniform temperature distribution over the length of the reactor and, therefore, enabled a precise temperature measurement by means of a thermocouple placed directly on the reactor tube.

After the surface reduction, a gas mixture with a ratio of H$_2$:CO$_2$ = 4:1 with He as carrier gas and a total flow of 10 mL$_n$/min was set on the mass flow controllers to the bypass of the reactor after closing the valves to the reactor. The gas flows and purities were: 6 mL$_n$/min and 99.995% for H$_2$; 1.5 mL$_n$/min and 99.998% for CO_2; 2.5 mL_n/min and 99.999% for He. After a stable gas mixture was achieved, the reaction gas stream was let through the reactor and the measurement was started. The reactor oven was then heated up at a rate of 2 K/min from 393 K to 823 K (oven set points). The effective measured temperatures were 390 K to 810 K on the reactor. The reaction was carried out at ambient pressure. Given the small amount of sample, eventual gaseous products arising from further reductions in phases in the material during the temperature ramp would be negligible, compared to the gas flow in the reactor.

A spectrum of masses from 1 to 100 u was measured approximatively every two minutes with an OmniStar Pfeiffer Mass Spectrometer (MS). The product analysis by means of MS is discussed in the next section.

For the analysis of the products of the catalytic reaction, we applied a semi-quantitative method by means of mass spectrometry (MS), which allowed for the comparison of the activity and selectivity among the catalysts. This method is a variation of the quantitative analysis method we developed and described in detail in our previous work [19,20]. In comparison to that method, we accounted for the pressure fluctuations over the MS capillary by normalizing the intensity of the signal to a reference pressure, which was defined and the same for all experiments. The internal pressure was logged for every spectrum; hence the normalization was applied on each spectrum. Since the signal intensity in the Faraday detector was linear to the incident ions, it was also directly proportional to the pressure difference over the capillary. Therefore, a normalized and comparable signal was obtained if the signal intensity (ion current) was multiplied by a correction factor f_{corr}, calculated from the logged MS partial pressure p_{MS} and the reference pressure p_{ref}:

$$f_{corr} = p_{MS}/p_{ref} \qquad (5)$$

where $p_{ref} = 10^{-3}$ Pa. After the integration of the peaks over one half of a mass/charge ratio m/z, the integration area was normalized with f_{corr}:

$$A_{corr,mz} = A_{mz}/f_{corr} \qquad (6)$$

where A_{mz} is the area of the integrated signal peak for one m/z. With this method, the (integrated) signal intensity for different catalysts could be directly compared for each m/z, allowing a semi-quantitative analysis. For the quantification of the partial pressures of the products with this method, a pressure-normalized calibration for all products would be required. For our purposes, this was not necessary.

In any case, the MS peaks of C1-C5 hydrocarbons (HCs) strongly overlapped with the peaks of CO_2 and CO. To identify the ideal m/z peaks for the analysis, reference electron ionization patterns for each compound (C1-C5 HC, MeOH, EtOH, CO, CO_2, H_2 and He) were plotted, based on the data reported on the NIST Chemistry WebBook [21]. Based on these data, the reference peaks were selected, as shown in Table S1.

3. Results

3.1. Structure, Morphology and Composition of Fe-Co Nanoalloys

The TEM images in Figure 1a–c show the morphology of the as-prepared nanoalloys. The NPs sizes follow a log-normal distribution, typical for this technique [22]. The mean NP size spans from 10 to 13 nm, with sample 30Fe70Co having the largest size. It is worth noting that the distributions in Figure 1d,f are, intrinsically, number-weighted distributions. However, in Table 1 we report the volume-weighted mean diameter (or de Brouckere mean $\bar{d}^{TEM}$), assuming spherical particles that can be directly compared to the mean crystallite determined by XRD.

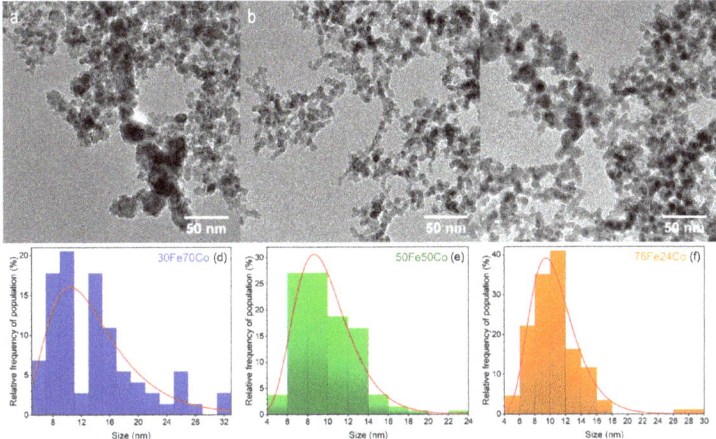

Figure 1. TEM images of xFe(100-x) Co samples with x = 30, 50, 76 in frames (**a**–**c**), respectively. The corresponding NPs size distributions are given in frames (**d**–**f**).

The EDX profiles in Figure 2d,f have been measured on Fe-Co nanoalloys along the red path, indicated in the HAADF-STEM images in Figure 2a–c, respectively. The red squares in the graphs show that the Co/(Co + Fe) ratio is homogeneous from the core to the surface within the single NPs for all samples.

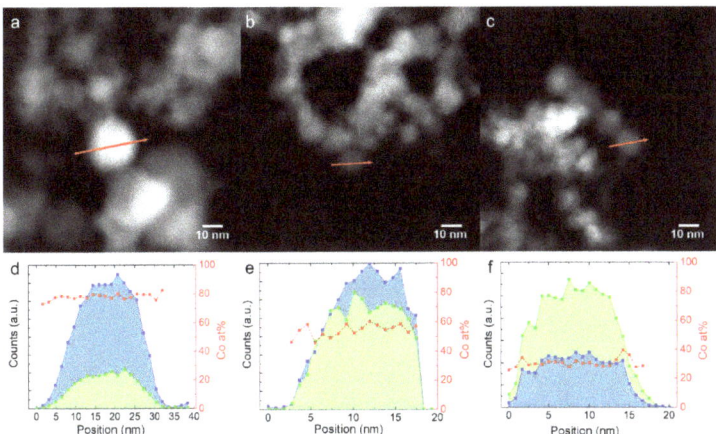

Figure 2. Fe and Co EDX composition profiles along the red arrow, shown in the STEM image above them. (**a**,**d**): 30Fe70Co; (**b**,**e**): 50Fe50Co; (**c**,**f**): 76Fe24Co. The blue and green profiles represent Co and Fe counts, respectively. Red squares represent the calculated Co atomic content.

The overall Fe and Co contents measured by SEM-EDX on the NPs batch are reported in Table 1. In the two samples with higher Fe content, the SEM-EDX agree with the nominal precursor composition within the experimental uncertainty. The sample 30Fe70Co seems to have a higher Fe content (30 ± 2 wt%) than its precursor. This discrepancy can be explained by the higher vapor pressure of Fe (37 Pa at 2000 K) with respect to Co (20 Pa at 2000 K) [23], which leads to a higher evaporation rate of Fe. It is, however, possible to calibrate the relative Fe content in the precursor in order to achieve the target nanoalloy composition. Such an approach requires that the activity coefficients of the two elements do not vary strongly with composition, otherwise the relative evaporation rates would change during the synthesis resulting in a non-homogeneous batch. Another option is to use two closely spaced and independently controlled evaporation sources to produce a vapor mixture with the target composition [24].

The specific surface area of each sample measured with the BET method, S_A^{BET}, is reported in Table 1. Assuming spherical and isolated NPs (no interfaces), it is possible to evaluate the specific surface area S_A^{TEM} from the TEM size distribution through the formula:

$$S_A^{TEM} = \frac{S_{TOT}}{\rho V_{TOT}} = \frac{6 \sum_i d_i^2}{\rho \sum_i d_i^3}, \quad (7)$$

where d_i is the diameter of the i-th particle observed by TEM and ρ is the density of the NPs. The results are compared to S_A^{BET} in Table 1. For the calculation, ρ is taken as a weighted average of the bulk densities of Fe, $\rho_{Fe} = 7960$ kg m^{-3}, and Co, $\rho_{Co} = 8830$ kg m^{-3}.

Figure 3 shows the XRD patterns of Fe and the three nanoalloys, all showing the Bragg reflections characteristic of a body-centered cubic (bcc) α-phase. The (110) peak shifts toward higher angles with increasing Co content, indicating the decrease in the lattice parameter, as shown in Table 1, compatible with the smaller atomic radius of Co. In sample 30Fe70Co only, traces of a second face-centered cubic (fcc) Fe-Co phase are detected. A Fe$_3$O$_4$ magnetite-like inverse spinel structure is also detected, featuring broader diffraction peaks corresponding to a crystallite size of about 3 nm. The relative abundance of this phase within the sample decreases with increasing Co content and is barely visible in the XRD pattern of 30Fe70Co.

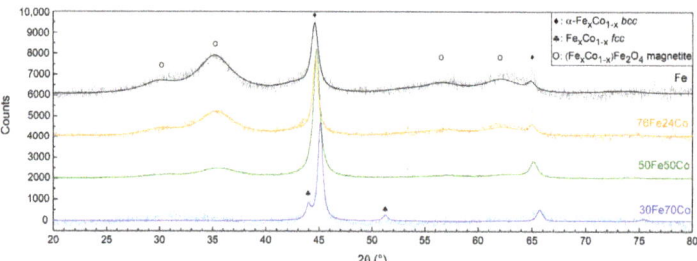

Figure 3. XRD patterns of the three Fe-Co nanoalloys compared to elemental Fe NPs. The peaks of the bcc Fe-Co alloy shift toward higher angles with increasing Co content.

3.2. Catalytic Properties of Fe-Co Nanoalloys

The catalytic performances of the samples are summarized in Figure 4a–e, showing the CO$_2$ conversion, CO yield and CH$_4$ yield for the nanoalloys and elemental NPs. Here the yield is defined as the ratio between the output flow of the product and the input flow of CO$_2$, meaning that the sum of all product yields is equal to the CO$_2$ conversion. A product-by-product comparison of the same results is presented in Figure S1 of the SI. Figure 4f shows the C$_2$-C$_5$ yield of the same samples. The Fe-Co nanoalloys all exhibit activity in the CO$_2$ hydrogenation reaction where CH$_4$ and CO are the major carbon containing products on all catalysts. Compared to elemental Fe or Co NPs, the nanoalloys show conversion yields for CO and CH$_4$ with a different temperature dependence and genuinely new activity towards the formation of C$_2$-C$_5$ hydrocarbons. The total CO$_2$ conversion is given as the sum of CO and CH$_4$ yields. As explained in the Materials and Methods section, the analysis of C$_2$-C$_5$ hydrocarbons is semi-quantitative but the absolute yield would be < 1%, therefore neglecting it does not significantly affect the CO$_2$ conversion curve. The formation of alcohols, such as methanol or ethanol, is not detected. The highest CO$_2$ conversion (at moderate temperatures) is 18%, achieved by 76Fe24Co at 662 K with 28% CH$_4$ and 72% CO selectivity. While the effects of the alloy composition on the reaction temperatures will be discussed later, no clear trend for the maximum conversion of C_2-C_5 products as a function of the Fe content in the alloy can be determined.

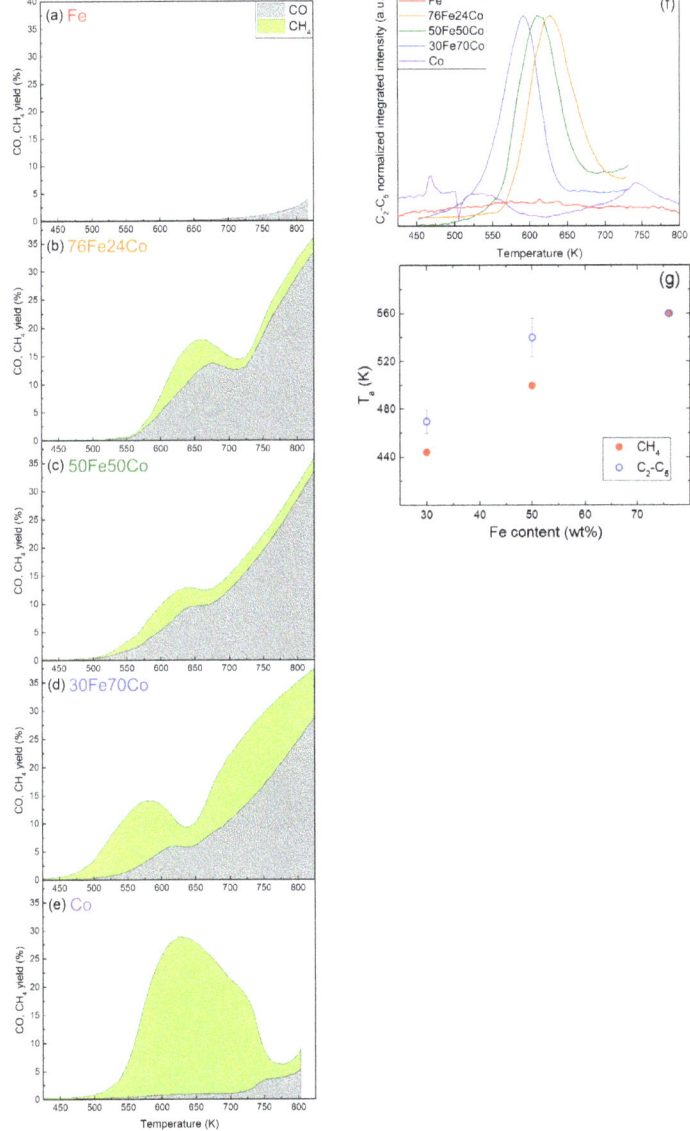

Figure 4. Catalytic properties of samples (**a**) Fe, (**b**) 76Fe24Co, (**c**) 50Fe50Co, (**d**) 30Fe70Co, (**e**) Co in a flow reactor with 4:1 H_2:CO_2 ratio, 1 bar and 10 mL$_n$ min^{-1}, measured by mass spectroscopy. The stacked area-filled plots show the total CO_2 conversion with the separated contributions of CO yield in grey and CH_4 yield in green; (**f**) conversion of CO_2 into C_2-C_5 HCs obtained semi-quantitatively by summing the normalized MS signals for m/z = 26, 29, 30, 39, 56, 57, and 70; (**g**) activation temperature T_a of the nanoalloy catalysts for CH_4 and C_2-C_5 production as function of their Fe content. The error bars represent the standard deviation in the activation temperatures of the different C_2-C_5 HCs, as detailed in Table S2.

Figure 4g reports the activation temperatures T_a for the formation of CH_4 and C_2-C_5 HCs products, as determined by analyzing the reference MS peaks for each molecule versus the Fe content in the nanoalloys. The error bar given on the data points for C_2-C_5 corresponds to the standard deviation of the T_a values observed for the various products, as shown in Table S2.

Two trends are observed: first, the formation of C_2-C_5 HCs starts at temperatures higher than those observed for CH_4, with the exception of 76Fe24Co, where $T_a = 560$ K for all HCs. Second, T_a increases with increasing Fe content in the alloy. This is also clearly visible in Figure 4a–e for CH_4 and in Figure 4f for C_2-C_5 products, where the peaks shift to higher temperatures with increasing Fe content. Table S2 in the SI lists the detailed results of the kinetic analysis for the three nanoalloys, reporting the kinetic reaction range, the temperature of maximum activity, and the activation energy E_a of each product.

As concerns E_a, in almost all cases, it is larger for the C_2-C_5 products than for CH_4, as listed in Table S2. Moreover, Figure S2 shows that the E_a values of the C_2-C_5 products increase with increasing Fe content, thus following a trend similar to T_a. The E_a values are in the ranges 65–80 kJ mol^{-1} for Fe30Co70, 80–140 kJ mol^{-1} for Fe50Co50, and 110–275 kJ mol^{-1} for Fe76Co24. It is worth noting that these values are measured at conditions constrained by the thermal stability of the materials and are meant to be compared within this work, not to other Fe-based catalysts that have been properly activated with thermal treatments that typically last several hours above 573 K.

4. Discussion

4.1. Structure, Morphology, and Composition of Fe-Co Nanoalloys

The volume-weighted average NPs size $\bar{d}^{TEM}$ calculated from the size distributions of Figure 1 under the assumption of spherical NPs is in good agreement with the average crystallite size, determined by XRD, as shown in Table 1, supporting the idea that NPs are mostly single crystalline. The specific surface area S_A^{TEM}, estimated from the TEM size distributions under the additional hypothesis of isolated NPs, is in good agreement with the BET result S_A^{BET} for the two samples, 76Fe24Co and 50Fe50Co, suggesting that NPs retain most of their free surface despite aggregation. Contrarily, for sample 30Fe70Co, S_A^{TEM} is about 50% larger than S_A^{BET}, indicating that aggregation leads to a loss of surface area in favor of interface area. This is in qualitative agreement with the presence of NPs in Figure 1a, the size of which largely exceeds the average value, hinting at inter-particle coalescence.

HAADF-STEM images, as shown in Figure 2a–c, show that the NPs are surrounded by a 2–3 nm thick shell with a lower contrast, which suggests that the average atomic number is lower than in the core. This information, combined with XRD, points to the metal core/oxide shell nature of the NPs. The oxide shell has a cubic inverse spinel structure (space group $Fd\bar{3}m$) and forms during the slow air exposure for XRD and TEM experiments [25]. The shell is identified as magnetite Fe_3O_4 for the elemental Fe NPs, and cobalt ferrite $(Fe_xCo_{1-x})_3O_4$ [26,27] for the Fe-Co nanoalloys. The STEM-EDX profiles in Figure 2e,f show that the Fe/Co ratio in the shell is similar to the core. The $(Fe_xCo_{1-x})_3O_4$ peaks decrease in intensity with increasing Co content, indicating a higher oxidation resistance, as already reported for nearly equimolar Fe-Co nanoalloys [28].

After the CO_2 hydrogenation experiments the oxide, is not detected anymore by XRD, as shown in Figure S3, nor is carbon laydown. The mean crystallite size is increased as expected after exposure to temperatures >800 K; however, the alloy is still observed with no phase segregation.

4.2. Catalytic Properties of Fe-Co Nanoalloys: Compositional Effects

The Fe-Co alloys, compared to elemental Fe and Co, display a much higher selectivity toward CO, an intermediate one for CH_4, and show activity in the catalytic formation of C_2-C_5 hydrocarbons, as shown in Figure 4f. The high activity of the Fe-Co alloys towards CO means that the surface of these catalysts is rich in adsorbed CO, which is not the case for Fe and Co, although for opposite reasons, as discussed later. The abundance of adsorbed CO suggests that the C_2-C_5 HCs are synthesized via RWGS + FT reactions (Equations (2) and (3)) rather than via direct CO_2 hydrogenation (Equation (4)).

Remember that Co acts as a purely Sabatier catalyst, while Fe is active towards the RWGS reaction, and only at high temperatures.

The theoretical work of Liu et al. [29] on the reduction of CO_2 on transition metal surfaces provides a deeper understanding of these results. They conclude that CO_2 is strongly adsorbed on Fe surfaces, but this is not favorable for CO_2 decomposition into CO + O, because the energy of the transition state between adsorption and decomposition causes a high reaction barrier. On the other side, CO_2 is weakly adsorbed on Co surfaces, but the transition state energy is similar to that of Fe, resulting in a lower decomposition barrier. A qualitative representation of the relevant energy levels and barriers is sketched in Figure 5. The adsorption and decomposition of the CO_2 molecule are necessarily the first steps in CO_2 hydrogenation and here we argue that they are key steps for the Fe-Co catalytic system. Based on the results of these calculations, it is possible to better interpret the catalytic activity measured for pure Fe and Co. Fe surfaces are covered with CO_2 that is easily adsorbed but is too strongly bound to react, showing very poor CO_2 conversion rates in general. When the temperature is high enough to overcome the CO_2 decomposition barrier, CO is the only product because it is thermodynamically favored. On Co surfaces, the weak adsorption energy means that the surface is not rich in adsorbed CO_2 but most of the adsorbed molecules decompose into CO because of the low decomposition energy barrier. The few and isolated CO molecules cannot then meet to start the HC chain by forming C-C bonds, so that the hydrogenation reaction necessarily goes on from CO to CH_4.

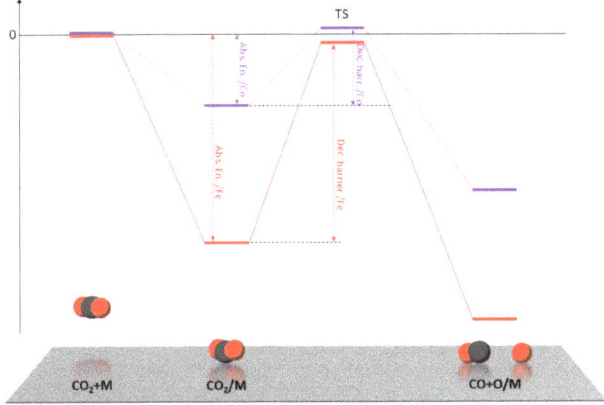

Figure 5. Schematic representation of the calculated energy levels of (from left to right) free CO_2, adsorbed CO_2, transition state, decomposed CO_2 into CO + O. In red, relative to Fe surface; in purple, relative to Co surface. The dashed vertical lines highlight the difference between Fe and Co in adsorption energy and decomposition barrier [29].

Although we are not aware of similar theoretical studies for Fe-Co surfaces, it is reasonable to expect a composition-dependent intermediate behavior for the alloy. This view is supported by our experimental finding that both E_a and T_a of HCs formation increase with increasing Fe content in the alloy. Higher E_a and T_a values mean a stronger interaction of the adsorbed species with the surface (i.e., a more Fe-like behavior). By adjusting the Fe/Co ratio in the alloy, it is therefore possible to tune the CO_2 adsorption energy and decomposition barrier in order to have more CO_2 adsorbed than on Co but, at the same time, an easier decomposition to CO than on Fe. The result is an abundance of adsorbed CO on the alloy surface, which is the cause for the higher selectivity to CO and the starting point in the formation of C-C bonds for the growth of HC chains in FT synthesis. The low C_2-C_5 yields (in Figure 4f the y-axis is in arbitrary units, but the yield is < 1%), the limitations of the MS technique, and the limited amount of data do not allow for a quantitative composition dependent analysis. It will be interesting to investigate this aspect in a future work and at higher operating pressures.

The T_a values of the various C_2-C_5 products are similar and slightly higher than for CH_4 (ΔT_a ~35 K for 30Fe70Co and 50Fe50Co). This difference in T_a is explained considering the superior kinetics of the Sabatier reaction compared to the RWGS + FT reaction [30], especially at lower temperatures, where the Sabatier reaction is also thermodynamically favored. With increasing temperature, CO synthesis through RWGS starts to compete with the Sabatier reaction. This is the reason why ΔT_a almost vanishes for the 76Fe24Co nanoalloy (i.e., the composition that shows the highest activation temperature $T_a = 560\ K$).

5. Conclusions

The Fe-Co nanoalloys synthesized by inert gas condensation exhibit enhanced catalytic activities for CO_2 hydrogenation compared to elemental Fe and Co NPs, being also slightly active toward the synthesis of C_2-C_5 hydrocarbons. On top of CO_2 conversion and product yield, thanks to the developed set-up, based on mass spectrometry, it is possible to measure the activation temperature T_a and estimate the activation energy E_a for each reaction product. The observed increase in both T_a and E_a with rising Fe content in the nanoalloys, as well as their activity in C_2-C_5 synthesis, is interpreted based on the idea of composition-dependent CO_2 adsorption energy and decomposition barrier. The balance between the high density of adsorbed stable CO_2, typical of Fe, and the low density of easily decomposed CO_2, typical of Co, leads to higher activity and C-C bond formation on the Fe-Co surface, which initiates the hydrocarbon chain. This work is relevant in assessing the catalytic properties of Fe-Co nanoalloys, ruling out the effects of supports, metal/support interfaces, and promoters.

Supplementary Materials: The following are available online at http://www.mdpi.com/2079-4991/10/7/1360/s1. Table S1: Reference table for the assignments of mass spectrometer peaks. Table S2. Activation temperature (T_a) of selected m/z mass spectrometer peaks, which are assigned to C_1-C_5 hydrocarbon products, the corresponding temperature of the maximum activity Tmax, the starting and ending temperature of the kinetically determined reaction range (T1 kin and T2 kin), and activation energy (E_a). %Tmax is an indicator that tells us if the reaction is solely limited by the reaction kinetics; it gives the ratio of T2 kin compared to Tmax. R^2 is assigned to the Arrhenius plots (inverse T1 kin to T2 kin vs. the natural logarithm of the normalized MS signal) on which the activation energy is determined. Figure S1: Catalytic properties of Fe, Co and Fe-Co NPs in a flow reactor with 4:1 H_2:CO_2 ratio, 1 bar and 10 mL$_n$ min^{-1}, measured by mass spectroscopy. (**a**) CO_2 conversion (**b**) CO yield, (**c**) CH_4 yield, (**d**) conversion curves of the summed up C2-C5 mass spectrometer normalized signals for m/z = 26, 29, 30, 39, 56, 57, 70. Figure S2: Activation energies of C2-C5 product formation as a function of the Fe content in the alloy precursor. The colors corresponding to the different products are listed in the legend, with the m/z ratio of the MS reference peak increasing from top to bottom. Figure S3 (a): XRD of the Fe-Co samples after the CO_2 hydrogenation experiments, background corrected; (b): a detail of the main Fe-Co peak shifted to higher angular positions with increasing Co content.

Author Contributions: Conceptualization and methodology, M.C., R.M., N.P., and L.P.; investigation, M.C., R.M., N.P., A.M., and K.Z.; project administration and resources, L.P. and A.Z.; writing—original draft preparation, M.C.; writing—review and editing, all authors; supervision, L.P. and A.Z. All authors have read and agreed to the published version of the manuscript.

Funding: This research was funded by Schweizerischer Nationalfonds zur Förderung der Wissenschaftlichen Forschung (Swiss National Science Foundation), grant number 200021_163010/1, and by the Swiss Competence Center for Energy Research - Heat and Electricity Storage (innosuisse).

Acknowledgments: Stéphane Voeffray and Robin Délèze are acknowledged for the construction of the reactor.

Conflicts of Interest: The authors declare no conflict of interest.

References

1. Graves, C.; Ebbesen, S.D.; Mogensen, M.; Lackner, K.S. Sustainable hydrocarbon fuels by recycling CO_2 and H_2O with renewable or nuclear energy. *Renew. Sustain. Energy Rev.* **2011**, *15*, 1–23. [CrossRef]
2. Züttel, A.; Mauron, P.; Kato, S.; Callini, E.; Holzer, M.; Huang, J. Storage of Renewable Energy by Reduction of CO_2 with Hydrogen. *Chim. Int. J. Chem.* **2015**, *69*, 264–268. [CrossRef] [PubMed]
3. Jiao, F.; Li, J.; Pan, X.; Xiao, J.; Li, H.; Ma, H.; Wei, M.; Pan, Y.; Zhou, Z.; Li, M.; et al. Selective conversion of syngas to light olefins. *Science* **2016**, *351*, 1065–1068. [CrossRef] [PubMed]

4. Gao, P.; Li, S.; Bu, X.; Dang, S.; Liu, Z.; Wang, H.; Zhong, L.; Qiu, M.; Yang, C.; Cai, J.; et al. Direct conversion of CO_2 into liquid fuels with high selectivity over a bifunctional catalyst. *Nat. Chem.* **2017**, *9*, 1019–1024. [CrossRef]
5. Wei, J.; Ge, Q.; Yao, R.; Wen, Z.; Fang, C.; Guo, L.; Xu, H.; Sun, J. Directly converting CO_2 into a gasoline fuel. *Nat. Commun.* **2017**, *8*, 15174. [CrossRef]
6. Luo, M.; Davis, B.H. Fischer-Tropsch synthesis: Group II alkali-earth metal promoted catalysts. *Appl. Catal. Gen.* **2003**, *246*, 171–181. [CrossRef]
7. Li, W.; Wang, H.; Jiang, X.; Zhu, J.; Liu, Z.; Guo, X.; Song, C. A short review of recent advances in CO_2 hydrogenation to hydrocarbons over heterogeneous catalysts. *RSC Adv.* **2018**, *8*, 7651–7669. [CrossRef]
8. Satthawong, R.; Koizumi, N.; Song, C.; Prasassarakich, P. Comparative Study on CO_2 Hydrogenation to Higher Hydrocarbons over Fe-Based Bimetallic Catalysts. *Top. Catal.* **2014**, *57*, 588–594. [CrossRef]
9. Boreriboon, N.; Jiang, X.; Song, C.; Prasassarakich, P. Fe-based bimetallic catalysts supported on TiO_2 for selective CO_2 hydrogenation to hydrocarbons. *J. CO_2 Util.* **2018**, *25*, 330–337. [CrossRef]
10. Satthawong, R.; Koizumi, N.; Song, C.; Prasassarakich, P. Bimetallic Fe–Co catalysts for CO_2 hydrogenation to higher hydrocarbons. *J. CO_2 Util.* **2013**, *3–4*, 102–106. [CrossRef]
11. Wang, W.; Wang, S.; Ma, X.; Gong, J. Recent advances in catalytic hydrogenation of carbon dioxide. *Chem. Soc. Rev.* **2011**, *40*, 3703. [CrossRef] [PubMed]
12. Venturi, F.; Calizzi, M.; Bals, S.; Perkisas, T.; Pasquini, L. Self-assembly of gas-phase synthesized magnesium nanoparticles on room temperature substrates. *Mater. Res. Express* **2015**, *2*, 15007. [CrossRef]
13. Calizzi, M.; Chericoni, D.; Jepsen, L.H.; Jensen, T.R.; Pasquini, L. Mg-Ti nanoparticles with superior kinetics for hydrogen storage. *Int. J. Hydrogen Energy* **2016**, *41*, 14447–14454. [CrossRef]
14. Callini, E.; Pasquini, L.; Piscopiello, E.; Montone, A.; Antisari, M.V.; Bonetti, E. Hydrogen sorption in Pd-decorated Mg-MgO core-shell nanoparticles. *Appl. Phys. Lett.* **2009**, *94*, 221905. [CrossRef]
15. Rossi, G.; Calizzi, M.; Di Cintio, V.; Magkos, S.; Amidani, L.; Pasquini, L.; Boscherini, F. Local Structure of V Dopants in TiO_2 Nanoparticles: X-ray Absorption Spectroscopy, Including Ab-Initio and Full Potential Simulations. *J. Phys. Chem. C* **2016**, *120*, 7457–7466. [CrossRef]
16. Patelli, N.; Calizzi, M.; Migliori, A.; Morandi, V.; Pasquini, L. Hydrogen Desorption Below 150 °C in MgH_2-TiH_2 Composite Nanoparticles: Equilibrium and Kinetic Properties. *J. Phys. Chem. C* **2017**, *121*, 11166–11177. [CrossRef]
17. Patelli, N.; Migliori, A.; Morandi, V.; Pasquini, L. One-Step Synthesis of Metal/Oxide Nanocomposites by Gas Phase Condensation. *Nanomaterials* **2019**, *9*, 219. [CrossRef]
18. Lutterotti, L.; Bortolotti, M.; Ischia, G.; Londarelli, I.; Wenk, H.-R. Rietveld texture analysis from diffraction images. *Z. Krist. Suppl.* **2007**, *26*, 125–130. [CrossRef]
19. Mutschler, R.; Luo, W.; Moioli, E.; Züttel, A. Fast real time and quantitative gas analysis method for the investigation of the CO_2 reduction reaction mechanism. *Rev. Sci. Instrum.* **2018**, *89*, 114102. [CrossRef]
20. Mutschler, R.; Moioli, E.; Luo, W.; Gallandat, N.; Züttel, A. CO_2 hydrogenation reaction over pristine Fe, Co, Ni, Cu and Al_2O_3 supported Ru: Comparison and determination of the activation energies. *J. Catal.* **2018**, *366*, 139–149. [CrossRef]
21. NIST Chemistry WebBook-SRD 69. Available online: https://webbook.nist.gov/chemistry/ (accessed on 15 February 2020).
22. Hahn, H. Gas phase synthesis of nanocrystalline materials. *Nanostruct. Mater.* **1997**, *9*, 3–12. [CrossRef]
23. Lide, D.R. *CRC Handbook of Chemistry and Physics*; CRC Press: Boca Raton, FL, USA, 2003; p. 3485. ISBN 978-1466571143.
24. Patelli, N.; Migliori, A.; Morandi, V.; Pasquini, L. Interfaces within biphasic nanoparticles give a boost to magnesium-based hydrogen storage. *Nano Energy* **2020**, *72*, 104654. [CrossRef]
25. Signorini, L.; Pasquini, L.; Savini, L.; Carboni, R.; Boscherini, F.; Bonetti, E.; Giglia, A.; Pedio, M.; Mahne, N.; Nannarone, S. Size-dependent oxidation in iron/iron oxide core-shell nanoparticles. *Phys. Rev. B Condens. Matter Mater. Phys.* **2003**, *68*, 195423. [CrossRef]
26. Nlebedim, I.C.; Moses, A.J.; Jiles, D.C. Non-stoichiometric cobalt ferrite, $Co_xFe_{3-x}O_4$ (x = 1.0 to 2.0): Structural, magnetic and magnetoelastic properties. *J. Magn. Magn. Mater.* **2013**, *343*, 49–54. [CrossRef]
27. Dippong, T.; Levei, E.A.; Diamandescu, L.; Bibicu, I.; Leostean, C.; Borodi, G.; Barbu Tudoran, L. Structural and magnetic properties of $Co_xFe_{3-x}O_4$ versus Co/Fe molar ratio. *J. Magn. Magn. Mater.* **2015**, *394*, 111–116. [CrossRef]

28. Klencsár, Z.; Németh, P.; Sándor, Z.; Horváth, T.; Sajó, I.E.; Mészáros, S.; Mantilla, J.; Coaquira, J.A.H.; Garg, V.K.; Kuzmann, E.; et al. Structure and magnetism of Fe–Co alloy nanoparticles. *J. Alloys Compd.* **2016**, *674*, 153–161. [CrossRef]
29. Liu, C.; Cundari, T.R.; Wilson, A.K. CO_2 Reduction on Transition Metal (Fe, Co, Ni, and Cu) Surfaces: In Comparison with Homogeneous Catalysis. *J. Phys. Chem. C* **2012**, *116*, 5681–5688. [CrossRef]
30. Willauer, H.D.; Ananth, R.; Olsen, M.T.; Drab, D.M.; Hardy, D.R.; Williams, F.W. Modeling and kinetic analysis of CO_2 hydrogenation using a Mn and K-promoted Fe catalyst in a fixed-bed reactor. *J. CO_2 Util* **2013**, *3–4*, 56–64. [CrossRef]

© 2020 by the authors. Licensee MDPI, Basel, Switzerland. This article is an open access article distributed under the terms and conditions of the Creative Commons Attribution (CC BY) license (http://creativecommons.org/licenses/by/4.0/).

Article

Turning Waste into Useful Products by Photocatalysis with Nanocrystalline TiO$_2$ Thin Films: Reductive Cleavage of Azo Bond in the Presence of Aqueous Formate

Michele Mazzanti, Stefano Caramori *, Marco Fogagnolo, Vito Cristino and Alessandra Molinari *

Dipartimento di Scienze Chimiche e Farmaceutiche, Università di Ferrara, Via Luigi Borsari 46, 44121 Ferrara, Italy; michele.mazzanti@unife.it (M.M.); marco.fogagnolo@unife.it (M.F.); vito.cristino@unife.it (V.C.)
* Correspondence: cte@unife.it (S.C.); alessandra.molinari@unife.it (A.M.)

Received: 9 October 2020; Accepted: 22 October 2020; Published: 28 October 2020

Abstract: UV-photoexcitation of TiO$_2$ in contact with aqueous solutions of azo dyes does not imply only its photocatalytic degradation, but the reaction fate of the dye depends on the experimental conditions. In fact, we demonstrate that the presence of sodium formate is the switch from a degradative pathway of the dye to its transformation into useful products. Laser flash photolysis experiments show that charge separation is extremely long lived in nanostructured TiO$_2$ thin films, making them suitable to drive both oxidation and reduction reactions. ESR spin trapping and photoluminescence experiments demonstrate that formate anions are very efficient in intercepting holes, thereby inhibiting OH radicals formation. Under these conditions, electrons promoted in the conduction band of TiO$_2$ and protons deriving from the oxidation of formate on photogenerated holes lead to the reductive cleavage of N=N bonds with formation and accumulation of reduced intermediates. Negative ion ESI–MS findings provide clear support to point out this new mechanism. This study provides a facile solution for realizing together wastewater purification and photocatalytic conversion of a waste (discharged dye) into useful products (such as sulfanilic acid used again for synthesis of new azo dyes). Moreover, the use of TiO$_2$ deposited on an FTO (Fluorine Tin Oxide) glass circumvents all the difficulties related to the use of slurries. The obtained photocatalyst is easy to handle and to recover and shows an excellent stability allowing complete recyclability.

Keywords: TiO$_2$; azo dye; wastewater treatment; photocatalysis; sodium formate

1. Introduction

Very large amounts of dyes are annually produced and used in different industries, including textile, leather and paper industries [1]. Approximately 50–70% of the dyes available on the market are azo compounds, some of which have been reported to be or are suspected to be human carcinogens [2,3].

In addition, considering that up to 20% of dyestuff is discharged directly into the environment, it appears immediately evident that azo dyes are hazardous pollutants of high impact.

Removal of colored pollutants is accomplished by traditional physical techniques (adsorption on activated carbon, ion exchange on resins, coagulation, etc.) [4–7]. Nevertheless, the organic dye is simply transferred from water to another phase, causing secondary pollution and requiring the regeneration of the adsorbent. On the contrary, advanced oxidation processes (AOP) are emerging, because they are able to disrupt the dye molecule through the action of generated OH radicals [8–14]. Among AOPs, TiO$_2$-based photocatalysis has been the subject of numerous investigations since illumination of TiO$_2$, dispersed in water containing a dissolved dye, usually led to almost complete decoloration of the solution [15–20]. Hydroxyl radical formation and their non-selective reaction with dyes have been supposed as the main degradation pathway. Although various analytical techniques (HPLC, GC–MS, LC–MS, ^{1}H NMR, FT-IR) are available for the detection of the intermediates, the application of photocatalytic procedures for remediation of textile wastewaters remains rather limited [19]. A main drawback is that information available on reaction mechanisms involved in the degradation of dyes is still scant, and mineralization of intermediates is slower than the degradation of the parent compound. Until now, total mineralization has been observed for the photocatalytic degradation of most of the azo dyes only at long irradiation periods [21–24].

In this paper, we report that two alternative photocatalytic processes based on TiO_2 can be operative in aqueous solutions containing an azo dye. Methyl orange (MO) and acid orange 7 (AO7) were chosen as representative azodyes. Specifically, the investigation of the fate of the photogenerated charges and the direct detection of OH radicals highlight that sodium formate is the switch between the two mechanisms. In fact, in its absence, we show several independent pieces of evidence that the known photodegradation of the dye occurs. However, the crucial presence of formate selectively transforms the azodye molecule into the corresponding anilines upon a reductive cleavage of its N=N bond. Transient measurements, ESR spin trapping, photoluminescence, C/C_0 vs. irradiation time and ESI–MS data are in agreement with the role proposed for formate.

Although it has already been reported that illuminated TiO_2 is able to reduce nitroaromatic molecules to the corresponding anilines [25–30], as far as we know, the one presented here is the first example concerning the possibility of reaching two simultaneous goals by photocatalysis, namely the removal of a waste from water by converting it into useful building blocks.

In addition, we circumvented all the difficulties related to the use of slurries of TiO_2, since we deposited the semiconductor oxide onto a TCO-glass substrate [31,32], obtaining a photocatalyst that is easy to handle and to recover, which shows excellent stability, allowing complete recyclability and operation at ambient temperature and atmospheric pressure.

2. Experimental

2.1. Materials

Methyl orange (MO, Carlo Erba, Milan, Italy, >99.98%) and acid orange 7 (AO7, VWR, Milan, Italy, ≥97%) were purchased and used without further purification.

All other reagents employed in the experiments were also commercial: HCOONa (Sigma, Milan, Italy, 99.5%), spectrophotometric grade ethanol (Fluka, Milan, Italy, >99.8%), coumarin, (Sigma, >99%) and the spin trap 5,5′-dimethylpyrroline N-oxide (DMPO, Sigma, ≥97%).

2.2. Preparation of TiO_2 on Glass Substrate

Conductive fluorine tin oxide (FTO) substrates (Pilkington TEC 7) were cleaned via sonication in 2-propanol for 10 min and dried under a warm air stream. A compact titania blocking underlayer was fabricated on top of the FTO by overnight hydrolysis of 0.4 M $TiCl_4$ drop cast on top of the FTO slides (10 × 1 cm^2), followed by firing in air at 450 °C for 30 min. Porous TiO_2 films were obtained on top of the blocking underlayer by doctor blading a commercial terpineol based paste (Dyesol 18 NRT), followed by sintering at 500 °C for 45 min in air. The geometrical TiO_2 film area, composed by anatase nanocrystals of the approximate size of 20 nm, was 1 cm^2, and the film thickness was about 7 microns.

2.3. Structural Characterization

Atomic force microscopy (AFM) images were collected using a Digital Instruments Nanoscope III scanning probe microscope (Veeco-Digital Instruments, Plainview, NY, USA). The instrument was equipped with a silicon tip (RTESP-300 Bruker, Billerica, MA, USA) and operated in tapping mode. Surface topographical analysis of AFM images was carried out with a NanoScope Analysis 1.5.

Scanning electron microscopy of the films was obtained with a JEOL JSM-7001F FEG-SEM (JEOL Ltd., Tokyo, Japan) scanning electron microscope (SEM) apparatus. Measurements were performed at 10.0–20.0 k eV electron beam energy, and the working distance was maintained between 3 and 10 mm. Surface morphology images were acquired in top-down and tilted modes, whereas cross-sectional analysis was performed putting the films on a 90° stub. X-ray diffraction (XRD) measurements were performed using a BRUKER D8 Advance X-ray diffractometer equipped with a Sol-X detector, working at 40 kV and 40 mA. The X-ray diffraction patterns were collected in a step-scanning mode with steps of $\Delta 2\theta = 0.02°$ and a counting time of 10 s/step using Cu Kα1 radiation (λ = 1.54056 Å) in the 2θ range of 3–80° using an incident grazing angle set-up. The crystal size (L) was estimated from the full width at half maximum (*FWHM*) of the intense (101) peak after correction from the instrumental broadening, by using the Scherrer Equation as follows:

$$L = \frac{B \times \lambda}{FWHM \times \cos\theta} \quad (1)$$

where B = 0.9, λ = 1.54056 Å and θ is the angle at which the peak maximum is observed.

2.4. Electrochemistry and Photoelectrochemistry

Open circuit chronopotentiometry to determine the quasi-Fermi level of TiO_2 was performed in a three electrode cell (FTO supported TiO_2 film/Pt/SCE (SCE = Standard Calomel Electrode) as follows: initially the TiO_2 photoanode was positively polarized in the dark at 0.8 V vs. SCE for 300 s and then allowed to reach a nearly steady potential in the dark. This equilibration process was slow and occurred on the time scale of several hundred seconds. Usually after 400 s the dark potential is acceptably stable, and AM 1.5 G light is shone on the TiO_2 substrate, causing generation of charge carriers (electron/hole couples), which may recombine or undergo separation and storage within the semiconductor. Owing to their strongly positive quasi-Fermi potential, holes are at least partially scavenged by the electrolyte before recombination occurs, allowing electrons to accumulate inside TiO_2 and causing a sudden negative drift of the photovoltage. Illumination of the photoelectrode is maintained until a steady state value of the photopotential is attained, the reading of which provides the electrochemical potential of TiO_2 vs. SCE. Upon restoration of the dark conditions, a fast (for the time scale of the experiment) decay of the photovoltage occurs owing to recombination. The same cell setup was used to record TiO_2 cyclic voltammetries (CV) in the dark in the absence and in the presence of MO. The redox properties of the MO dye were investigated by cyclic voltammetry at a glassy carbon electrode, and potentials were referred against SCE.

2.5. Quantum Chemical Computation

DFT (Density Functional Theory) and TDDFT (Time Dependent Density Functional Theory) calculations were carried out with Gaussian 09A2 at the B3LYP 6–311 g,d level of theory [33]. The MO structure was pre-optimized at the PM6 level in vacuo before running the geometry optimization via DFT B3LYP 6-311 g,d in water solvent, described within the polarizable continuum model (PCM) approximation [34]. TDDFT computation was carried out on the optimized MO geometry, by considering the 10 lowest vertical singlet excitations in the presence of water (PCM).

2.6. Laser Spectroscopy

Transient absorption spectra were obtained under ns excitation of the third harmonic (355 nm) of a Nd/Yag Q switched laser oscillator described elsewhere [35]. The laser beam was diverged with a plano concave lens to achieve an energy density of 5 mJ/cm^2 at the surface of a TiO_2/FTO thin film, held at 45 degrees with respect to the laser beam. The thin film was in contact with an aqueous phase inside a spectrophotometric cell having an internal volume of ca. 3 mL. The aqueous phase consisted either of pure water or 0.1 M formate solution, in agreement with the explored photocatalytic conditions.

The aqueous solutions could be purged with argon, if needed. The probe light was monochromatized (Applied Photophysics, Leatherhead, UK)) and passed through the sample; then, it was focused into the Acton triple grating monochromator (50 lines/mm) and fed to an R3896 photomultiplier biased at 450 V. A stack of two 380 nm cut off filters placed in front of the Acton monochromator prevented laser stray light from reaching the photomultiplier. ΔA vs. time traces were collected with a Teledyne LeCroy Waverunner 604Zi oscilloscope (Teledyne Technologies, Thousand Oaks, CA, USA) having an input impedance of 1 MΩ, and transferred to a PC with custom built software, which also controlled synchronization of the spectrometer elements. An acceptable S/N ratio was achieved by averaging 100 laser shots at each sampled wavelength.

2.7. ESR Spin Trapping

ESR spin trapping experiments were carried out with a Bruker ER200 MRD spectrometer equipped with a TE201 resonator (microwave frequency of 9.4 GHz). The samples were deaerated aqueous suspensions of TiO_2 (Evonik) containing 5,5'-dimethylpyrroline N-oxide (DMPO, 5×10^{-2} M) as spin trap and sodium formate (0.1 M) when requested. The deaerated suspensions were transferred into a flat quartz cell inside a box under N_2 atmosphere and directly irradiated (Hg medium pressure lamp with a cut off filter, $\lambda \geq 360$ nm) in the ESR cavity. No signals were obtained in the dark or during irradiation of the solution in the absence of TiO_2.

2.8. Photoluminescence Experiments

The fluorescence measurements ($\lambda_{exc} = 332$ nm and $\lambda_{emiss} = 455$ nm) were performed at room temperature with a Jobin Yvon Spex Fluoromax II spectrofluorimeter equipped with a Hamamatsu R3896 photomultiplier. Both emission and excitation slits were set at 5.0 nm during the measurements. For this kind of experiment, the FTO/TiO_2 slide was put inside a Pyrex tube containing a deaerated aqueous solution (3 mL) of the dye (MO or AO7, $C_0 = 10$ ppm), HCOONa (0.1 M, when requested) and coumarin (1×10^{-4} M) and then irradiated ($\lambda \geq 360$ nm, 4 h). After irradiation, the fluorescence spectrum of 7-hydroxycoumarin, eventually formed, was recorded.

2.9. Prolonged Irradiations

Prolonged irradiations were carried out with a Helios Italquartz Q400 medium-pressure Hg lamp, Milan, IT. A glass cut-off filter was used ($\lambda \geq 360$ nm). The incident flux was 2.75×10^{16} photons s^{-1} cm^{-2}, calculated from the measured radiant power density in mW cm^{-2} [36]. UV–vis spectra were recorded with a Jasco V-630 double beam spectrophotometer. In a typical experiment, an FTO/TiO_2 sheet was put inside a Pyrex tube of 15 mL capacity in front of the unique optical face, and a volume (3 mL) of an aqueous solution containing the dye (MO or AO7, $C_0 = 10$ ppm) and, when requested, HCOONa (0.1 M) was added. The pH of this solution was 7.1. The closed test tube was firstly filled with N_2 by bubbling the gas for 20 min, and then the FTO/TiO_2 sheet was back irradiated for the desired period with the external Hg lamp. At the end of irradiation, the sample was transferred into a spectrophotometric cuvette, and the electronic spectrum was recorded. From the decrease in the absorption maximum of the band characteristic of this dye, the concentration of the remaining dye was evaluated, and a decay curve reporting C/C_0 vs. irradiation time was built. Each experiment was repeated three times in order to evaluate the error. Blank experiments were run both in the dark, but in the presence of TiO_2, and also illuminating the solution containing the dye (as described above), but in the absence of TiO_2.

For recycle experiments, the just employed FTO/TiO_2 sheet was thoroughly washed with water and dried in air at room temperature. Then it was used again in a subsequent experiment.

2.10. ESI–MS Investigation

Mass spectra were recorded using a LCQ Duo (ThermoQuest, San Jose, CA, USA), equipped with an electrospray ionization source (ESI), monitoring the precursor-to-product ion transitions of m/z 100 to 400 in negative ionization mode. A sample prepared as described in Section 2.6 was 4 h irradiated. Then the solution was analyzed. Ethanol (10% *v/v*) was used in place of HCOONa in order to have a nonionic hole scavenger.

3. Results and Discussion

3.1. Structural Properties

Figure 1A shows the cross sectional SEM imaging of a typical titania thin film used for the present work. The film displayed a porous nanocrystalline nature with a quite homogeneous thickness around 7 µm, composed of sintered particles having a size in the order of few tens of nanometers, with frequent randomly distributed larger aggregates. SEM imaging, even at 50,000× magnification (Figure 1B) did not allow each single anatase nanoparticle to be clearly resolved, which were better shown by tapping mode AFM maps (1 µm × 1 µm scanning area), where we could appreciate single particles having an approximate size of 20–30 nm together with larger aggregates resulting in lumps of 50–100 nm diameter, homogeneously distributed within the film. The sintered particles left pores and cavities which made the film permeable to the electrolyte. XRD analysis (Figure 1D) confirmed the presence of the anatase polymorph (tetragonal cell with a = b = 3.78 Å, c = 9.51 Å and 90° angles) with major reflections from (101), (004) and (200) planes. The estimation of the crystalline size (L) from the Scherrer equation provided a value of 19.8 nm, in agreement with the smallest particles observed from the scanning probe microscopies and with the nominal particle size (18 nm) contained in the commercial 18 NRT paste, confirming that each particle was a coherent scattering domain composed by single anatase crystal.

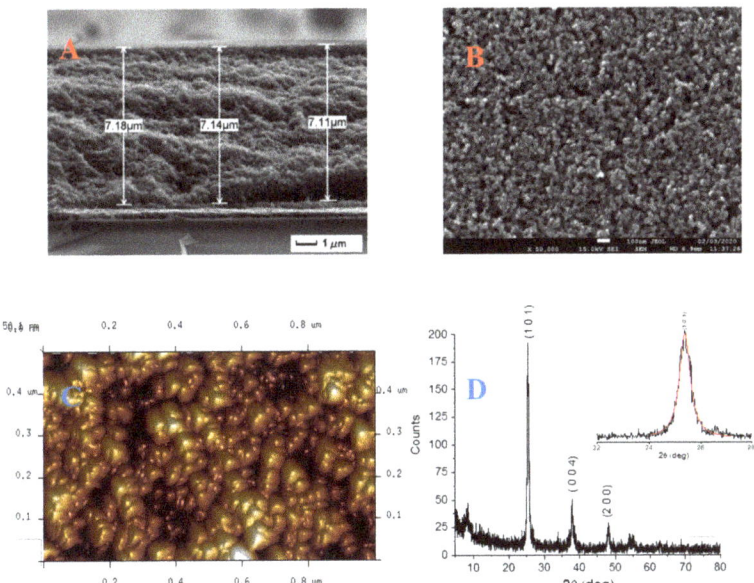

Figure 1. (**A**) Cross sectional SEM view of a nanocrystalline TiO_2 thin film deposited on FTO equipped with a compact TiO_2 blocking underlayer; (**B**) top view of the TiO_2 surface at 50,000× magnification; (**C**) AFM imaging of the mesoporous film surface. (**D**) XRD of the thin film showing the main diffraction peak of anatase. The inset shows the 101 peak fitted with a Lorentzian function to extract the FWHM for crystal size analysis.

3.2. Energetics

The energetics relevant to the photocatalytic process were explored by combining electrochemical, photoelectrochemical and quantum chemical calculations. The quasi-Fermi levels of TiO_2 in the presence of either 0.1 M Na_2SO_4 or HCOONa are shown in Figure 2, together with the redox levels of MO associated to the HOMO and LUMO isodensity surfaces obtained from TDDFT calculations at the 6311 G, d level in the presence of water (PCM). The quasi-Fermi potential of TiO_2 was obtained via open circuit chronopotentiometry under AM 1.5 G illumination (Figure S1), whereas the redox levels of the dye were obtained by cyclic voltammetry at a glassy carbon electrode (Figure S2). MO was characterized by irreversible oxidation and reduction processes, which appeared as diffusion limited waves having comparable intensity and peaking at −0.80 and +0.72 V vs. SCE in 0.1 M sodium formate supporting electrolyte. In sodium sulfate these processes maintained the same general features observed in HCOONa, with a slightly increased electrochemical gap (E^{peak}_{RED} = −0.9 vs. SCE; E^{peak}_{OX} = +0.74 V vs. SCE). While these processes were irreversible, meaning that chemical changes followed the charge transfer reaction, we could assume that the primary event involved the charge transfer to and from the electrode from and to HOMO and LUMO orbitals of the organic dye. We observed that the HOMO was a π orbital with a significant contribution of the donor dimethyl-amino group, whereas the lowest π* orbital with antibonding properties (LUMO) exhibited a major contribution from the diazo (–N=N–) group. The reliability of the calculation was corroborated by the fair match of the computed vertical transitions with the experimental spectrum in water (Figure S3). The contribution of the diazo group to the LUMO suggests that electrochemical reduction of MO may indeed result in the localized cleavage of this molecular unit. MO reduction became thermodynamically feasible with illuminated TiO_2 in the presence of sodium formate. Under such conditions, efficient hole scavenging by $HCOO^-$ allowed electron build up within the semiconductor, resulting in a large negative drift (absolute amplitude ca. 1.2 V) of its electrochemical potential (E_F) up to −0.85 V vs. SCE, matching well the reduction peak of MO in the same electrolyte (Figure S4). CVs reported in Figure S5 indicated that under steady illumination, most of the trap states (oxygen vacancies) below the conduction band edge, evidenced by a pre-wave starting at ca. −0.5 V vs. SCE and peaking at ca. −0.7/−0.8 V vs. SCE (Figure S5), may have become occupied and acted as an electron reservoir for the intended reduction process. By contrast, in sodium sulfate, hole scavenging by water remained in competition with charge recombination. Hence the TiO_2 quasi-Fermi level under AM 1.5 G illumination only reached −0.45 V vs. SCE, barely at the onset of the reduction processes of Figure S6, leading to an insufficient energy overlap between the TiO_2 reducing states and the electron acceptor levels to drive an efficient reductive photocatalysis.

The direct clear observation of the dark electrocatalytic reaction between the TiO_2 electrode and MO in solution was, however, quite elusive, owing to a combination of factors which include (i) the large (super)capacitive response of the porous electrode; (ii) the low solubility of MO in water (10^{-3}M); and (iii) slow (for the time scale of cyclic voltammetry) electron transfer kinetics. Only at 10 mV/s did the cyclic voltammetry of the TiO_2 film in the presence of 10^{-3} M MO show a weak intermediate wave, which could be assigned to MO reduction at the TiO_2 surface (Figure S5).

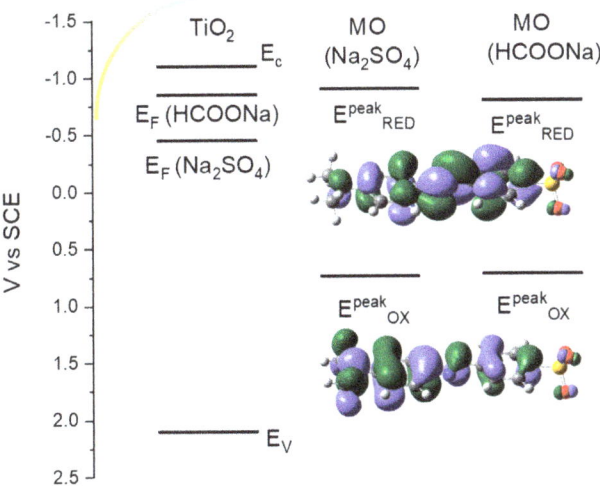

Figure 2. Energy levels (represented on the electrochemical scale vs. SCE) involved in the reductive degradation of the methyl orange (MO) dye at TiO_2 photoelectrodes. The exponential Density of States (DOS) of the TiO_2 is represented by the yellow-to-cyan curve starting at ca. −0.55 V vs. SCE, consistent with the CV results reported in the Supporting Information. In the presence of formate, the electrochemical potential of TiO_2 matches well the first reduction peak potential of MO in the same solvent.

3.3. Transient Spectra of TiO_2 Thin Films

After having obtained an insight on the interfacial energetics involved in the photocatalytic reduction of MO, we explored the recombination kinetics with optical transient spectroscopy. Transient spectra of TiO_2 thin films, collected in the 0.1 ms–s timeframe, were related to the simultaneous presence of trapped charge carriers having opposite sign (electrons and holes). In the presence of pure water, the spectra showed a stronger absorption in the blue visible region, which decreased by moving to the red (Figure 3A). The overall spectral shape was maintained throughout the whole timeframe, spanning two orders of magnitude, from 0.25 ms to 0.033 s, during which a general decrease of the amplitude of the transient signals was observed, according to a multiexponential decay (Figure S7). The half-life ($t_{1/2}$) of the decays in both the blue and red parts of the visible region were comparable, about 0.033 s (0.036 s at 420 nm, 0.032 s at 720 nm). According to Bahnemann et al. [37] we assigned the blue absorption to trapped holes which could result in the formation of surface bound OH radicals. Pulse radiolytic experiments have confirmed that the •OH radical absorption rises in the UV region and extends in the visible up to 470 nm [38], while earlier experiments by laser flash photolysis [39] pointed out the 420 nm absorption that we also observed as a shoulder in the spectrum reported in Figure 3A.

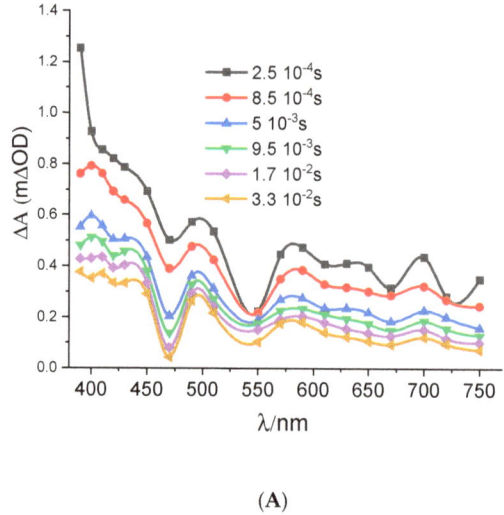

(A)

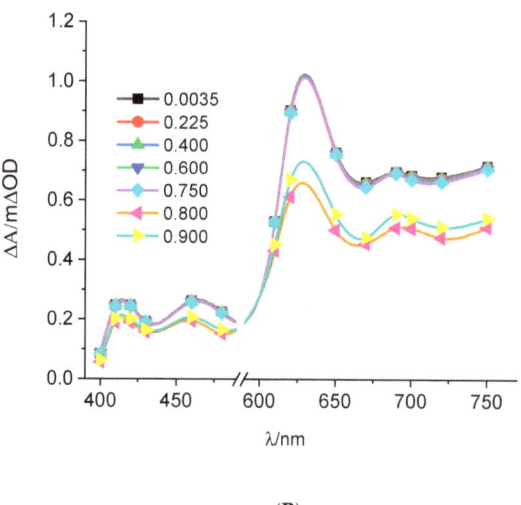

(B)

Figure 3. Transient spectra of TiO$_2$ thin films in contact with aqueous solutions under aerated conditions. Laser excitation wavelength 355 nm, 5 mJ/cm^2/pulse. (**A**) Pure water: difference spectra were sampled at delays from 0.25 ms to 0.33 s. (**B**) Aqueous solution of formate (0.1 M): difference spectra were sampled at delays from 3.5 ms to 0.9 s.

The broad long wavelength features were instead assigned to absorption of trapped electrons, the half-life of which nearly doubled in the absence of oxygen, which acted as an electron scavenger, affording a t$_{1/2}$ in the order of 0.07 s, as evidenced from the 720 nm kinetics reported in Figure S8. Upon addition of 0.1 M sodium formate, the absorption of trapped holes was reduced by a factor of ca. 90%, whereas the electron absorption was concomitantly increased and longer lived (Figure 3B). The strong absorption band between 600 and 700 nm is consistent with reports by Bahnemann et al. [37] in the presence of other holes scavengers like polyvinyl alcohol. It is interesting to observe different electron absorption features by moving to pure water to 0.1 M formate. In particular, the strong absorption centered at 630 nm was dominating in the presence of formate, while it was much less evident in pure water. Such a feature is consistent with previous research which assigns the optical transitions of trapped electrons as originating from $T_{2g} \rightarrow E_g$ states of Ti(III), the coordination sphere of which may reflect the presence of chelating formate anions in our case [40].

3.4. ESR Spin Trapping Experiments

Photoexcitation of deaerated suspensions of TiO_2 powder in water containing the spin trap DMPO caused the formation of a quartet, 1:2:2:1 ($a_N = a_H$ = 14.5 G), ascribable to the paramagnetic adduct [DMPO-OH]$^\bullet$ in accordance with previous investigation [41,42] (Figure 4A). Hydroxyl radicals, formed by the reaction between water on the TiO_2 surface and positive holes, were then trapped by DMPO, according to reactions (R1) and (R2):

$$H_2O + h^+ \rightarrow H^+ + OH^\bullet + e^- \tag{R1}$$

$$DMPO + OH^\bullet \rightarrow [DMPO - OH]^\bullet \tag{R2}$$

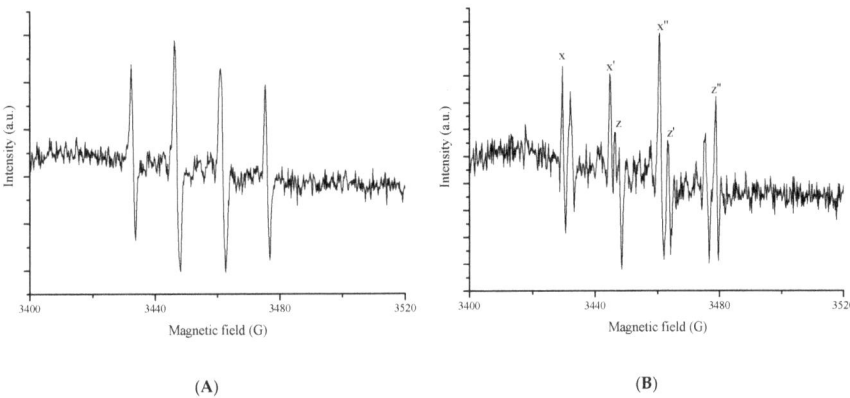

Figure 4. ESR spin trapping spectra obtained upon photoexcitation of TiO_2 powder suspended in deaerated water solutions containing the spin trap DMPO (5 ×10^{-2} M) (part (**A**)) and also sodium formate (0.1 M, part (**B**)).

Interestingly, when the experiment was carried out in the presence of aqueous formate (0.1 M), the signal of the paramagnetic adduct [DMPO-OH] was no longer observed and instead a new triplet of doublets was obtained (Figure 4B). Its hyperfine splitting constants (a_N (x-x') = 15.4 G, a_H (x-z) = 18.5 G) were consistent with the formation and trapping of a formate radical ($CO_2^{-\bullet}$), as shown in reaction (R3) [43,44]:

$$HCOO^- + h^+ \rightarrow H^+ + CO_2^{-\bullet} \tag{R3}$$

3.5. Photoluminescence Experiments

The $^\bullet$OH radical formation was investigated also by using coumarin as a fluorescent probe. In fact, the reaction between coumarin and $^\bullet$OH radicals produced 7-hydroxycoumarin (among the possible hydroxylated products), which is a strongly luminescent compound, according to reaction (R4):

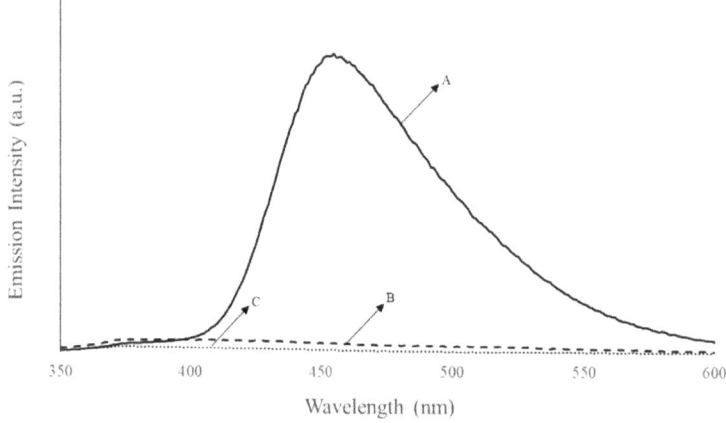

(R4)

This method has been successfully applied for the detection of hydroxyl radicals generated by photoexcited TiO_2 [45,46]. Therefore, the FTO/TiO_2 sheet immersed in a deaerated aqueous solution containing MO (10 ppm) and coumarin (1×10^{-4} M) was irradiated ($\lambda > 360$ nm, 4 h) both in the absence and in the presence of formate (0.1 M), and the emission spectrum of the irradiated solution was recorded. It was observed (Figure 5) that an important emission from 7-hydroxycoumarin was obtained in the absence of HCOONa (curve A), indicating the occurrence of reaction (R4). Conversely, in the presence of formate, fluorescence was obtained (curve B), evidencing that OH radical formation was negligible, in accordance with ESR spin trapping results. Curve B was in fact similar to the spectrum of the starting solution before irradiation, which contained unreacted coumarin (curve C).

Figure 5. Emission spectrum of 7-hydroxycoumarin obtained upon illumination ($\lambda > 360$ nm) of FTO/TiO_2 sheet immersed in deaerated aqueous solutions containing coumarin (1×10^{-4}M), (A); (B) is the spectrum obtained in the same conditions as A, except that 0.1 M HCOONa was present (curve B). (C) refers to the emission of the sample before illumination.

All the previous results lead to common conclusions that are summarized in the following: transient spectra confirm that charge separation is extremely long lived in nanostructured TiO_2 thin films, making them suitable to drive both oxidation and reduction reactions by exploiting the respective trapped charge carriers. Formate anions are very efficient in intercepting holes, inhibiting OH radical formation and allowing both the concentration and lifetime of trapped electrons to increase in order to use them for multi-electron reductive photocatalysis. On the basis of these considerations, we have experimental evidence that photoexcitation of TiO_2 immersed in water can lead to selective reduction processes if formate is used as a hole scavenger. Moreover, protons formed as described in reaction (R3) and electrons promoted in the conduction band could perform hydrogenation reactions. In the following, we explore this possibility considering reductive cleavage of N=N bonds in MO and in AO7, chosen as representative azo dyes.

3.6. Prolonged Irradiation Experiments

An FTO/TiO$_2$ slide, having a TiO$_2$ coated area of 1 cm^2, immersed in a deaerated aqueous solution containing MO (C$_0$ = 10 ppm) and HCOONa (0.1 M), was irradiated from the FTO side ($\lambda \geq 360$ nm). This illumination geometry is referred to as back illumination mode. During irradiation, a loss of the solution color that corresponded to a decrease of the absorption band of MO (maximum at 464 nm), having a charge transfer (CT) character, was observed. Figure 6 reports a C/C$_0$ ratio of MO as a function of irradiation time. It was observed that after 240 min irradiation, about 80% of the starting dye had disappeared from the solution (full squares). Control experiments pointed out that this result was exclusively ascribable to photocatalytic activity of TiO$_2$, since irradiation of the dye solution in the absence of semiconductor led to no spectral variations (i.e., no photolysis, Figure 6 full triangles). Moreover, no spectral variation was observed when the nanocrystalline TiO$_2$ thin film was put in contact with the dye solution and kept in the dark.

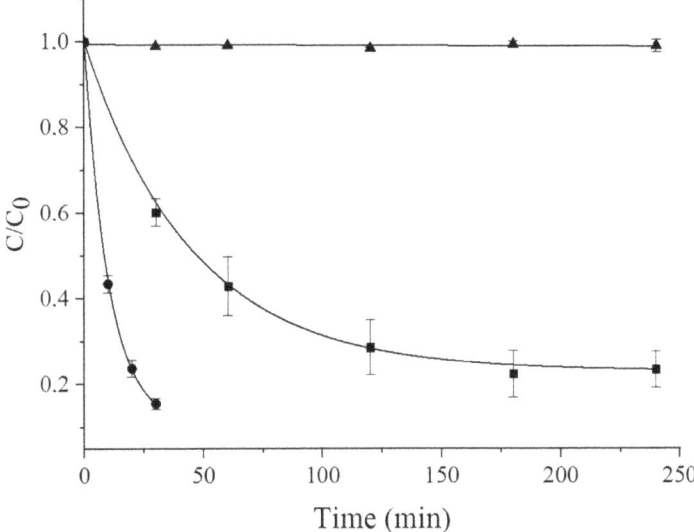

Figure 6. C/C$_0$ ratio vs. time profiles obtained upon irradiation ($\lambda \geq 360$ nm) of an FTO/TiO$_2$ slide (active TiO$_2$ geometrical area =1 cm^2) immersed in deaerated aqueous solutions containing HCOONa (0.1 M) and MO (full squares) or AO7 (full circles). C$_0$ = 10 ppm. Full triangles: the same solutions irradiated in the absence of FTO/TiO$_2$.

An analogous behavior was observed when AO7 was used instead of MO. C/C$_0$ vs. time reported in Figure 6 (full circles) showed that around 90% of this dye disappeared from the solution after only 30 min irradiation.

Curves similar to those of Figure 6 were also obtained in the absence of formate (not shown here), where formed OH radicals were responsible for the oxidative degradation of the dye. Thus, monitoring the disappearance rate of the target dye is not the most appropriate way to establish the nature of the photocatalytic reaction. It should be taken into account that fading of the dye solution is only an indication of the interruption of conjugation.

With the aim of ascertaining that the observed decrease of dye concentration is due to the reductive cleavage of the azo bond with formation of the two amine derivatives, we recorded the negative ion mode ESI–MS spectrum of the irradiated solution (Figure 7). Besides the m/z peak at 304 relative to the residual MO in its anionic form, the dominant base peak at m/z 172 corresponded to the anionic form of sulfanilic acid [47]. A similar result was obtained in the case of AO7 (Figure S9).

Conversely, no evidence of the m/z 172 peak was obtained in the ESI–MS spectrum recorded after photocatalytic processing of the dye in the absence of formate (Figure S10). Detection of a peak with m/z 227 was in agreement with the formation of a byproduct whose molecular weight was consistent with hydroxyl group substitution on a phenyl ring [18]. In addition, many examples in the literature report that oxidation of aromatic organic molecules (such as drugs, pesticides) proceeds by the addition of hydroxyl radicals before ring breaking [15,21,48,49]. Thus, the pattern reported in Figure S4 is characteristic of a degradative process promoted by OH radicals.

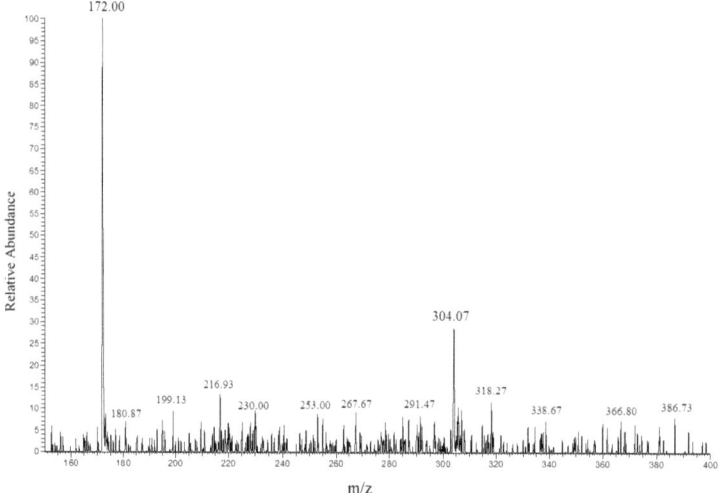

Figure 7. ESI–MS spectrum of the deaerated solution containing MO (10 ppm) and ethanol (10% *v/v*) after irradiation in the presence of FTO supported TiO_2 (1 cm^2). Ethanol is used here in place of formate as a non-ionic hole scavenger.

These results give evidence that when OH production is suppressed, the two azodyes MO and AO7 undergo a hydrogenation reaction on the diazo N=N bond (via protons and electrons addition). The subsequent cleavage of this bond leads to the formation of the two corresponding amine fragments, N, N-dimethyl 1, 4-phenylendiamine (from MO), 1-amino, 2-napthol (from AO7) and sulfanilic acid (from both MO and AO7), as summarized in Scheme 1.

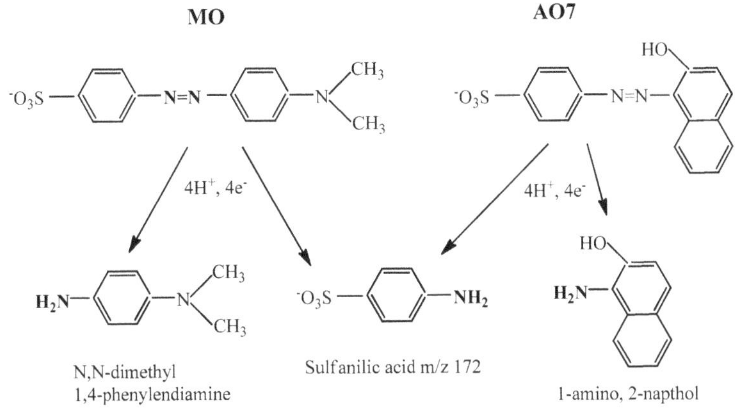

Scheme 1. Representation of N=N reductive cleavage of MO and of AO7 with formation of amine intermediates and of sulfanilic acid.

The photocatalytic cleavage of the diazo bond, here observed, most likely occurred at the surface of the anatase mesoporous film, with the involvement of protons provided by the oxidation of either formate or ethanol, as a result of photo-hole scavenging and of electrons stored in long living trap states inside the TiO_2 nanoparticles, and clear evidence of electron accumulation within the TiO_2 thin films was provided by both photoelectrochemical and spectroscopic means. Electron accumulation inside TiO_2 was also evident to the naked eye when using TiO_2 P25 (10 mg) nanoparticles suspended in a deaerated aqueous solution containing HCOONa (0.1 M) and irradiated with UV light ($\lambda \geq 360$ nm). While photogenerated holes were efficiently scavenged by formate, electrons, in the absence of any acceptor, accumulated, causing the formation of Ti(III) centers, responsible for the grey–blue color of the overall suspension [50]. After this was achieved, MO was added to the blue suspension through a septum. In the following minutes, the blue color bleached and a simultaneous decrease of the absorption in the 400–500 nm region was observed in the UV–visible spectrum of the solution after centrifugation (data not shown). This result confirms the ability of MO to scavenge electrons.

The alternative proton source could be water oxidation by photogenerated TiO_2 holes (reaction (R5)):

$$H_2O + h^+ \rightarrow H^+ + OH^\bullet + e^- \tag{R5}$$

However, since hydroxyl radical formation was completely inhibited in the presence of formate (see Sections 3.3 and 3.4), we can conclude that it is formate or another organic hole scavenger, like ethanol, that provides the required protons that participate in the reductive process schematized below.

This new synthetic route occurring in water by photoexcited TiO_2 is of particular significance; in fact, aqueous TiO_2 photocatalysis that is usually considered an AOP useful in depollution converts here an azo dye into useful reduced intermediates, one of which, sulfanilic acid, could be directly recycled in the synthetic process of new azo dye. Moreover, HCOONa oxidation leaves no residue (it is converted to CO_2) in the reaction environment, and the whole photocatalytic process is carried out at pH around 7 with no adjustments.

3.7. Recycle

Recyclability of the photocatalyst was evaluated using the same FTO/TiO_2 slide in several consecutive photocatalytic experiments. Typically, the FTO/TiO_2 immersed in a deaerated aqueous solution containing MO (10 ppm) and HCOONa (0.1 M) is back irradiated ($\lambda \geq 360$ nm) for a period of 4 h. At the end of the illumination period, MO disappearance (that we established to be due to reductive cleavage of N=N bond) is spectrophotometrically evaluated. The slide is then washed with distilled water and reused in a subsequent experiment. Figure 8 reports the amount of dye (%) disappeared in five consecutive runs. It can be noted that, within the experimental error, no decrease in photocatalytic performance was observed. This result indicates that the studied photocatalytic system is perfectly stable and completely re-usable.

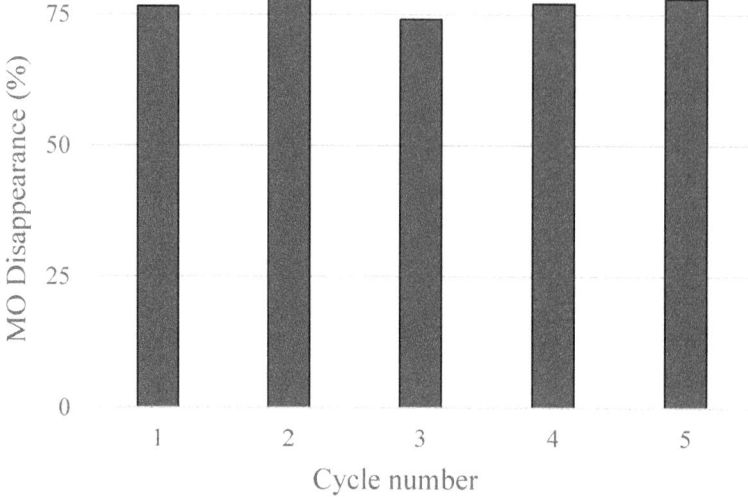

Figure 8. MO disappearance (%) obtained by irradiating (4 h, λ ≥ 360 nm) the same FTO supported TiO_2 thin film during five consecutive photocatalytic runs.

Moreover, three different FTO/TiO_2 slides (all prepared with the procedure described in 2.2) were used for the same photocatalytic experiment and gave very similar results, demonstrating that also the preparation procedure provides reproducible photocatalysts.

4. Conclusions

This study demonstrates that the fate of azo dyes dissolved in aqueous solution in the presence of illuminated TiO_2 can be determined by the experimental conditions. In fact, the usual photocatalytic degradation of the dye by OH radicals can be inhibited by the presence of sodium formate or of other hole scavengers that are able to release protons upon oxidation.

Laser flash photolysis experiments show that charge separation is extremely long lived in nanostructured TiO_2 thin films, making them suitable to drive both oxidation and reduction reactions. When a suitable hole scavenger, like formate, is present, holes are consumed in the sub-ms time scale, consistent with ESR spin trapping and photoluminescence experiments, demonstrating the formation of OH radicals. ESI–MS analysis of irradiated samples shows that multi-electron reduction of the azo dye on photoexcited TiO_2 is operative. In particular, a reductive cleavage of N=N bond of the dye takes place, with formation of aromatic anilines and of sulfanilic acid, which in turn can be recovered and reused for the synthesis of new dyestuffs. Formate is also the source of required protons and its oxidation leaves no residue in the solution.

This result is of interest because wastewaters, colored by the presence of dyes, can be the starting material for the conversion of pollutants into useful products by photoexcited TiO_2 at ambient temperature and atmospheric pressure. In addition, immobilization of TiO_2 on FTO glasses gives a heterogeneous system that is very stable, easy to separate from the solution medium, completely recyclable for countless times and with high reproducibility in the preparation. For this, the findings presented here may provide a new highly efficient and low cost method for azo dye wastewater treatment, opening to a proper circular approach.

Supplementary Materials: The following are available online at http://www.mdpi.com/2079-4991/10/11/2147/s1, Figure S1: Open circuit chronopotentiometry, Figure S2: Cyclic voltammetry of MO at a glassy carbon electrode, Figure S3: Experimental vs computed MO spectrum, Figure S4: Cyclic voltammetry of TiO_2 in HCOONa, Figure S5: Cyclic voltammetry of TiO_2 in Na_2SO_4, Figure S6: Cyclic voltammetry of MO at a TiO_2 electrode, Figure S7: 430 nm kinetics in pure water, Figure S8: 720 nm decay kinetics, Figure S9: ESI-MS spectrum in AO7 (10 ppm) and ethanol (10% v/v), Figure S10: ESI-MS spectrum in MO.

Author Contributions: S.C. and V.C., investigation, methodology and writing; M.M., M.F., A.M, investigation, data curation, writing; A.M., supervision, review). All authors have read and agreed to the published version of the manuscript.

Funding: The research was funded by the H2020 Interreg Program Italy-Croatia PEPSEA project (ID number 10047424) (S.C. and V.C.).

Acknowledgments: A special acknowledgement goes to Roberto Argazzi for his help in photoluminescence measurements. Nicola Grassi, who gained a Bachelor degree at the University of Ferrara, is acknowledged for performing some experiments. We thank Tatiana Bernardi for her support in ESI–MS measurements, Nicola Bazzanella and Michele Orlandi (UniTN) for SEM imaging and Gabriele Bertocchi for small angle XRD measurements. Finally, we kindly acknowledge the University of Ferrara for financial support (Fondi FAR 2018) and the Interreg Italy–Croatia PEPSEA (ID number 10047424) project.

Conflicts of Interest: The authors declare no conflict of interests.

References

1. Zollinger, H. *Color Chemistry: Synthesis, Properties and Applications of Organic Dyes and Pigments*, 2nd ed.; VCH: Vancouver, BC, Canada, 1991.
2. Gomes da Silva, C.; Faria, J.L. Photochemical and photocatalytic degradation of an azo dye in aqueous solution by UV irradiation. *J. Photochem. Photobiol. A Chem.* **2003**, *155*, 133–143. [CrossRef]
3. Brown, M.A.; De Vito, S.C. Predicting azo dye toxicity. *Crit. Rev. Environ. Sci. Technol.* **1993**, *23*, 249–324. [CrossRef]
4. Tang, W.Z.; An, H. UV/TiO_2 photocatalytic oxidation of commercial dyes in aqueous solutions. *Chemosphere* **1995**, *31*, 4157–4170. [CrossRef]
5. Meshko, V.; Markovska, L.; Mincheva, M.; Rodrigues, A.E. Adsorption of basic dyes on granular activated carbon and natural zeolite. *Water Res.* **2001**, *35*, 3357–3366. [CrossRef]
6. Kuo, W.S.; Ho, P.H. Solar photocatalytic decolorization of methylene blue in water. *Chemosphere* **2001**, *45*, 77–83. [CrossRef]
7. Galindo, C.; Jacques, P.; Kalt, A. Photooxidation of the phenylazonaphthol AO20 on TiO_2: Kinetic and mechanistic investigations. *Chemosphere* **2001**, *45*, 997–1005. [CrossRef]
8. Kuo, W.G. Decolorizing dye wastewater with Fenton's reagent. *Water Res.* **1992**, *26*, 881–886. [CrossRef]
9. Balanosky, E.; Fernadez, J.; Kiwi, J.; Lopez, A. Degradation of membrane concentrates of the textile industry by Fenton like reactions in iron-free solutions at biocompatible pH values (pH approximate to 7–8). *Water Sci. Technol.* **1999**, *40*, 417–424. [CrossRef]
10. Feng, W.; Nansheng, D.; Yuegang, Z. Discoloration of dye solutions induced by solar photolysis of ferrioxalate in aqueous solutions. *Chemosphere* **1999**, *39*, 2079–2085. [CrossRef]
11. Bandara, J.; Morrison, C.; Kiwi, J.; Pulgarin, C.; Peringer, P. Degradation/decoloration of concentrated solutions of Orange II. Kinetics and quantum yield for sunlight induced reactions via Fenton type reagents. *J. Photochem. Photobiol. A Chem.* **1996**, *99*, 57–66. [CrossRef]
12. Kang, S.F.; Liao, C.H.; Po, S.T. Decolorization of textile wastewater by photo-fenton oxidation technology. *Chemosphere* **2000**, *41*, 1287–1294. [CrossRef]
13. Arslan, I.; Akmehmet, T.; Tuhkamen, T. Advanced Oxidation of Synthetic Dyehouse Effluent by O_3, H_2O_2/O_3 and H_2O_2/UV Processes. *Environ. Technol.* **1999**, *20*, 921–931. [CrossRef]
14. Ince, N.H.; Gonenc, D.T. Treatability of a Textile Azo Dye by UV/H_2O_2. *Environ. Technol.* **1997**, *18*, 179–185. [CrossRef]
15. Vinodgopal, K.; Kamat, P. Combine electrochemistry with photocatalysis. *Chemtech* **1996**, *26*, 18–22.

16. Lizama, C.; Yeber, M.C.; Freer, J.; Baeza, J.; Mansilla, H.D. Reactive dyes decolouration by TiO_2 photo-assisted catalysis. *Water Sci. Technol.* **2001**, *44*, 197–203. [CrossRef] [PubMed]
17. Khataee, A.R.; Kasiri, M.B. Photocatalytic degradation of organic dyes in the presence of nanostructured titanium dioxide: Influence of the chemical structure of dyes. *J. Mol. Catal. A Chem.* **2010**, *328*, 8–26. [CrossRef]
18. Comparelli, R.; Fanizza, E.; Curri, M.L.; Cozzoli, P.D.; Mascolo, G.; Passino, R.; Agostiano, A. Photocatalytic degradation of azo dyes by organic-capped anatase TiO_2 nanocrystals immobilized onto substrates. *Appl. Catal. B Environ.* **2005**, *55*, 81–91. [CrossRef]
19. Konstantinou, I.K.; Albanis, T.A. TiO_2-assisted photocatalytic degradation of azo dyes in aqueous solution: Kinetic and mechanistic investigations: A review. *Appl. Catal. B Environ.* **2004**, *49*, 1–14. [CrossRef]
20. Pant, B.; Ojha, G.P.; Kuk, Y.-S.; Kwon, O.H.; Park, Y.W.; Park, M. Synthesis and Characterization of ZnO-TiO_2/Carbon Fiber Composite with Enhanced Photocatalytic Properties. *Nanomaterials* **2020**, *10*, 1960. [CrossRef]
21. Stylidi, M.; Kondarides, D.I.; Verykios, X.E. Pathways of solar light-induced photocatalytic degradation of azo dyes in aqueous TiO_2 suspensions. *Appl. Catal. B Environ.* **2003**, *40*, 271–286. [CrossRef]
22. Tanaka, K.; Padermpole, K.; Hisanaga, T. Photocatalytic degradation of commercial azo dyes. *Water Res.* **2000**, *34*, 327–333. [CrossRef]
23. Lachheb, H.; Puzenat, E.; Houas, A.; Ksibi, M.; Elaoui, E.; Guillard, G.; Hermann, J.M. Photocatalytic degradation of various types of dyes (Alizarin S, Crocein Orange G, Methyl Red, Congo Red, Methylene Blue) in water by UV-irradiated titania. *Appl. Catal. B: Environ.* **2002**, *39*, 75–90. [CrossRef]
24. Gouvea, C.A.K.; Wypych, F.; Moraes, S.G.; Duran, N.; Nagata, N.; Peralta-Zamora, P. Semiconductor-assisted photocatalytic degradation of reactive dyes in aqueous solution. *Chemosphere* **2000**, *40*, 433–440. [CrossRef]
25. Palmisano, G.; Garcia-Lopez, E.; Marcì, G.; Loddo, V.; Yurdakal, S.; Augugliaro, V.; Palmisano, L. Advances in selective conversions by heterogeneous photocatalysis. *Chem. Commun.* **2010**, *46*, 7074–7089. [CrossRef]
26. Kou, J.; Lu, C.; Wang, J.; Chen, Y.; Xu, Z.; Varma, R.S. Selectivity Enhancement in Heterogeneous Photocatalytic Transformations. *Chem. Rev.* **2017**, *117*, 1445–1514. [CrossRef] [PubMed]
27. Shiraishi, Y.; Hirakawa, H.; Togawa, Y.; Sugano, Y.; Ichikawa, S.; Hirai, T. Rutile Crystallites Isolated from Degussa (Evonik) P25 TiO_2: Highly Efficient Photocatalyst for Chemoselective Hydrogenation of Nitroaromatics. *ACS Catal.* **2013**, *3*, 2318–2326. [CrossRef]
28. Imamura, K.; Yoshikawa, T.; Hashimoto, K.; Kominami, H. Stoichiometric production of aminobenzenes and ketones by photocatalytic reduction of nitrobenzenes in secondary alcoholic suspension of titanium(IV) oxide under metal-free conditions. *App. Catal. B Environ.* **2013**, *134–135*, 193–197. [CrossRef]
29. Molinari, A.; Maldotti, A.; Amadelli, R. Probing the Role of Surface Energetics of Electrons and their Accumulation in Photoreduction Processes on TiO_2. *Chem. Eur. J.* **2014**, *20*, 7759–7765. [CrossRef]
30. Molinari, A.; Mazzanti, M.; Fogagnolo, M. Photocatalytic Selective Reduction by TiO_2 of 5-Nitrosalicylic Acid Ethyl Ester: A Mild Route to Mesalazine. *Catal. Lett.* **2020**, *150*, 1072–1080. [CrossRef]
31. Schneider, J.; Matsuoka, M.; Takeuchi, M.; Zhang, J.; Horiuchi, Y.; Anpo, M.; Bahnemann, D.W. Understanding TiO_2 photocatalysis: Mechanisms and materials. *Chem. Rev.* **2014**, *114*, 9919–9986. [CrossRef]
32. Pant, B.; Park, M.; Park, S.J. Recent Advances in TiO2 Films Prepared by Sol-Gel Methods for Photocatalytic Degradation of Organic Pollutants and Antibacterial Activities. *Coatings* **2019**, *9*, 613. [CrossRef]
33. Frisch, M.J.; Trucks, G.W.; Schlegel, H.B.; Scuseria, G.E.; Robb, M.A.; Cheeseman, J.R.; Scalmani, G.; Barone, V.; Mennucci, B.; Petersson, G.A.; et al. Gaussian Inc.: Wallingford, CT, USA, 2016.
34. Cossi, M.; Barone, V.; Cammi, R.; Tomasi, J. Ab initio study of solvated molecules: A new implementation of the polarizable continuum model. *Chem. Phys. Lett.* **1996**, *255*, 327–335. [CrossRef]
35. Ronconi, F.; Santoni, M.P.; Nastasi, F.; Bruno, G.; Argazzi, R.; Berardi, S.; Caramori, S.; Bignozzi, C.A.; Campagna, S. Charge injection into nanostructured TiO_2 electrodes from the photogenerated reduced form of a new Ru(II) polypyridine compound: The "anti-biomimetic" mechanism at work. *Dalton Trans.* **2016**, *45*, 14109–14123. [CrossRef] [PubMed]
36. Molinari, A.; Maldotti, A.; Amadelli, R. Effect of the electrolyte cations on photoinduced charge transfer at TiO_2. *Catal. Today* **2017**, *281*, 71–77. [CrossRef]
37. Bahnemann, D.W.; Henglein, A.; Lilie, J.; Spanhel, L. Flash photolysis observation of the absorption spectra of trapped positive holes and electrons in colloidal titanium dioxide. *J. Phys. Chem.* **1984**, *88*, 709–711. [CrossRef]
38. Lawless, D.; Serpone, N.; Meisel, D. Role of hydroxyl radicals and trapped holes in photocatalysis. A pulse radiolysis study. *J. Phys. Chem.* **1991**, *95*, 5166–5170. [CrossRef]

39. Bahnemann, D.W.; Henglein, A.; Spanhel, L. Detection of the intermediates of colloidal TiO$_2$ catalyzed photoreactions. *Faraday Discuss.* **1984**, *78*, 151–163. [CrossRef]
40. Bahnemann, D.W.; Hilgendorff, M.; Memming, R. Charge Carrier Dynamics at TiO$_2$ Particles: Reactivity of Free and Trapped Holes. *J. Phys. Chem. B* **1997**, *101*, 4265–4275. [CrossRef]
41. Molinari, A.; Argazzi, R.; Maldotti, A. Photocatalysis with Na$_4$W$_{10}$O$_{32}$ in water system: Formation and reactivity of OH radicals. *J. Mol. Catal. A Chem.* **2013**, *372*, 23–28. [CrossRef]
42. Maldotti, A.; Amadelli, R.; Carassiti, V.; Molinari, A. Catalytic oxygenation of cyclohexane by photoexcited (nBu$_4$N)$_4$W$_{10}$O$_{32}$: The role of radicals. *Inorg. Chim. Acta* **1997**, *256*, 309–312. [CrossRef]
43. Buettner, G.R. Spin trapping: ESR parameters of spin adducts. *Free Rad. Biol. Med.* **1987**, *3*, 259–303. [PubMed]
44. Molinari, A.; Samiolo, L.; Amadelli, R. EPR spin trapping evidence of radical intermediates in the photo-reduction of bicarbonate/CO$_2$ in TiO$_2$ aqueous suspensions. *Photochem. Photobiol. Sci.* **2015**, *14*, 1039–1046.
45. Czili, H.; Horvath, A. Applicability of coumarin for detecting and measuring hydroxyl radicals generated by photoexcitation of TiO$_2$ nanoparticles. *Appl. Catal. B Environ.* **2008**, *81*, 295–302.
46. Zhang, J.; Nosaka, Y. Quantitative Detection of OH Radicals for Investigating the Reaction Mechanism of Various Visible-Light TiO$_2$ Photocatalysts in Aqueous Suspension. *J. Phys. Chem. C* **2013**, *117*, 1383–1391.
47. Xie, S.; Huang, P.; Kruzic, J.J.; Zeng, X.; Qian, H. A highly efficient degradation mechanism of methyl orange using Fe-based metallic glass powders. *Sci. Rep.* **2010**, *6*, 21947. [CrossRef]
48. Molinari, A.; Sarti, E.; Marchetti, N.; Pasti, L. Degradation of emerging concern contaminants in water by heterogeneous photocatalysis with Na$_4$W$_{10}$O$_{32}$. *Appl. Catalys. B Environ.* **2017**, *203*, 9–17.
49. Pasti, L.; Sarti, E.; Martucci, A.; Marchetti, N.; Stevanin, C.; Molinari, A. An advanced oxidation process by photoexcited heterogeneous sodium decatungstate for the degradation of drugs present in aqueous environment. *Appl. Catal. B Environ.* **2018**, *239*, 345–351.
50. Vinodgopal, K.; Bedja, I.; Hotchandani, S.; Kamat, P.V. A Photocatalytic Approach for the Reductive Decolorization of Textile Azo Dyes in Colloidal Semiconductor Suspensions. *Langmuir* **1994**, *10*, 1767–1771.

Publisher's Note: MDPI stays neutral with regard to jurisdictional claims in published maps and institutional affiliations.

© 2020 by the authors. Licensee MDPI, Basel, Switzerland. This article is an open access article distributed under the terms and conditions of the Creative Commons Attribution (CC BY) license (http://creativecommons.org/licenses/by/4.0/).

Perspective

Single Particle Approaches to Plasmon-Driven Catalysis

Ruben F. Hamans [1,2], Rifat Kamarudheen [1] and Andrea Baldi [1,2,*]

1. Dutch Institute for Fundamental Energy Research (DIFFER), De Zaale 20, 5612 AJ Eindhoven, The Netherlands; r.f.hamans@vu.nl (R.F.H.); r.kamarudheen@differ.nl (R.K.)
2. Department of Physics and Astronomy, Vrije Universiteit Amsterdam, De Boelelaan 1081, 1081 HV Amsterdam, The Netherlands
* Correspondence: a.baldi@vu.nl

Received: 26 October 2020; Accepted: 20 November 2020; Published: 29 November 2020

Abstract: Plasmonic nanoparticles have recently emerged as a promising platform for photocatalysis thanks to their ability to efficiently harvest and convert light into highly energetic charge carriers and heat. The catalytic properties of metallic nanoparticles, however, are typically measured in ensemble experiments. These measurements, while providing statistically significant information, often mask the intrinsic heterogeneity of the catalyst particles and their individual dynamic behavior. For this reason, single particle approaches are now emerging as a powerful tool to unveil the structure-function relationship of plasmonic nanocatalysts. In this Perspective, we highlight two such techniques based on far-field optical microscopy: surface-enhanced Raman spectroscopy and super-resolution fluorescence microscopy. We first discuss their working principles and then show how they are applied to the in-situ study of catalysis and photocatalysis on single plasmonic nanoparticles. To conclude, we provide our vision on how these techniques can be further applied to tackle current open questions in the field of plasmonic chemistry.

Keywords: plasmonics; heterogeneous catalysis; photocatalysis; nanoparticles; single molecule localization; super-resolution microscopy; surface-enhanced Raman spectroscopy

1. Introduction

Localized surface plasmon resonances (LSPRs) arise from the coherent oscillation of free electrons upon illumination of metal nanoparticles. These resonances give rise to very large absorption and scattering cross sections, making plasmonic nanoparticles suitable for light harvesting applications such as photocatalysis [1–6], solar fuels generation [7–9], and integration in photonic devices [10–12]. For example, LSPRs have been used to increase the selectivity of propylene epoxidation [13], carbon dioxide reduction [14–16], and to increase the rate of H_2 dissociation [17] and ammonia decomposition [18]. Despite these demonstrations of enhanced catalytic selectivity and activity, a detailed understanding of the underlying mechanism is still lacking [18–28].

Unraveling the working mechanism of plasmonic catalysts can drive their rational design for specific applications. These studies are often performed at the ensemble level, resulting in measurements that are averaged over many particles and many active sites. The catalytic properties of metal nanoparticles, however, are intrinsically heterogeneous because of multiple factors: (i) even monodisperse nanoparticle ensembles present different reaction pathways for the same catalytic conversion [29], (ii) catalytic activity depends on the available surface facets [30] and the presence of defects [31,32], both of which can vary from particle to particle, and (iii) the structure of a catalyst changes during operation, resulting in spatiotemporal variations in catalytic activity [33–35].

Furthermore, in ensemble experiments it is typically difficult to characterize the weak signature of reaction intermediates that are crucial to unveil the reaction pathway [36].

To address these challenges, several techniques have emerged, allowing the study of the spatial and temporal heterogeneity of individual plasmonic nanoparticles and their photocatalytic reaction products under in-situ conditions [37]. Dark-field optical microscopy [38] and transmission electron microscopy (TEM) [33,35] have been used to follow the evolution of the optical and structural properties of active plasmonic photocatalysts with nanoscale spatial resolution. Mass spectrometry [39,40], infrared microscopy [37,39], and X-ray spectroscopy [41], on the other hand, while blind to the properties of the catalytic particles, allow the detection of the reaction products during the photocatalytic process. Furthermore, near-field techniques, such as infrared scanning near-field optical microscopy [42,43], tip-enhanced Raman spectroscopy [44,45], and scanning electrochemical microscopy [46,47] allow the interrogation of individual catalytic nanoparticles and obtain information on both their structure and catalytic performance. These techniques, however, typically have limited temporal (dark-field optical microscopy) or spatial (mass spectrometry) resolution and sensitivity, or require the use of expensive specialized equipment (TEM, infrared microscopy, near-field microscopy, and X-ray spectroscopy).

In this perspective, we highlight two experimental techniques that can be implemented in plasmonics research labs using conventional far-field optical microscopes and that are capable of detecting reaction products and intermediates both in real time and with nanometer spatial resolution: surface-enhanced Raman spectroscopy (SERS) and super-resolution fluorescence microscopy. SERS gives information on which intermediates and reaction products are present on the nanoparticle surface, but lacks sub-particle spatial resolution. Conversely, super-resolution fluorescence microscopy is able to characterize chemical reactions with nanometer resolution, but does not give any chemical information. We first describe the optical properties of metallic nanoparticles and the mechanisms by which they can enhance the rate of a chemical reaction. We then discuss the working principles of the two techniques and highlight selected articles that describe key observations made using these techniques in the context of plasmonic chemistry. Finally, we propose how these techniques can be further applied to study current open questions in plasmon-driven catalysis. For comprehensive reviews on SERS and super-resolution fluorescence microscopy in the context of catalysis, we refer to already existing literature [48–53].

2. Optical Properties of Metallic Nanoparticles

Let us consider the simple case of a metallic nanoparticle immersed in a harmonically oscillating electric field. The easiest case that can be solved analytically is that of a spherical nanoparticle with a radius much smaller than the oscillation wavelength. In this quasi-static approximation, the phase of the electric field is constant over the entire volume of the nanoparticle, and the problem can be simplified into that of a particle in an electrostatic field. This approximation allows us to rewrite Maxwell's equations into the Laplace equation, which, in the geometry considered here, allows us to define the polarizability α of the sphere as [54]:

$$\alpha = 4\pi R^3 \frac{\epsilon - \epsilon_m}{\epsilon + 2\epsilon_m}, \quad (1)$$

where R is radius of the sphere, ϵ is the wavelength-dependent complex permittivity of the sphere, and ϵ_m is the permittivity of the surrounding medium. The scattering (σ_{sca}) and absorption (σ_{abs}) cross sections can be calculated from the polarizability using [54]:

$$\sigma_{\text{sca}} = \frac{k^4}{6\pi} |\alpha|^2, \quad (2)$$

$$\sigma_{\text{abs}} = k \operatorname{Im}[\alpha], \quad (3)$$

where $k = 2\pi/\lambda$ is the wavevector. From the above equations it is clear that the particle polarizability, and therefore its scattering and absorption cross sections, diverge when $\epsilon = -2\epsilon_m$. For typical plasmonic metals such as gold and silver in vacuum ($\epsilon_m = 1$) or water ($\epsilon_m = 1.777$) this condition is satisfied in the visible part of the electromagnetic spectrum (Figure 1a). The magnitude of α remains finite due to the non-vanishing imaginary part of the permittivity ϵ, which accounts for losses in the metal (Figure 1b).

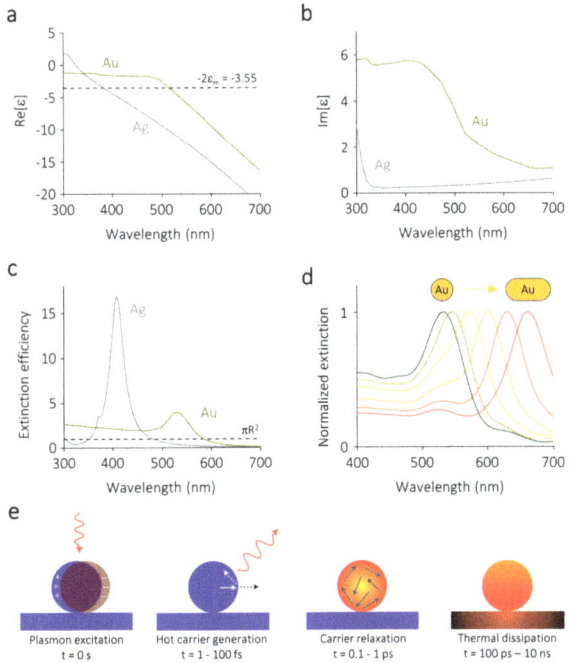

Figure 1. Optical properties of metallic nanoparticles. (**a**) Real and (**b**) imaginary part of the wavelength-dependent permittivities of gold (Au) and silver (Ag). The dotted line in panel (**a**) denotes the resonance condition $\epsilon = -2\epsilon_m$ when the surrounding medium is water. The permittivities are taken from the literature [55,56]. (**c**) Extinction cross sections of spherical nanoparticles of gold and silver with radii $R = 25$ nm. The cross sections are calculated using Mie theory and are normalized to the geometrical cross section πR^2, which is denoted by the dashed horizontal line. (**d**) Normalized extinction cross section of a gold sphere of $R = 25$ nm, that is elongated in steps of 10 nm to a gold rod with a total length of 100 nm. The cross sections are calculated using a finite-difference time-domain method. (**e**) Localized surface plasmon resonances (LSPRs) can decay radiatively (scattering) into photons or non-radiatively (absorption) into non-equilibrium charge carriers. These carriers relax via electron-electron scattering, followed by electron-phonon scattering, which heats up the nanoparticle and eventually also the surrounding medium. Panel (**e**) reproduced with permission of [2]. Copyright Nature Publishing Group, 2015.

For larger spheres the electric field phase variation inside the nanoparticle needs to be taken into account. If we consider a plane wave impinging on a spherical particle of arbitrary size in a homogeneous medium, this problem can still be treated analytically thanks to its spherical symmetry, resulting in the so-called Mie theory [54,57]. The analytical solution to Mie theory yields the electric fields inside and outside the particle, as well as its scattering and absorption cross sections. As shown in Figure 1c, the resulting extinction cross section ($\sigma_{\text{ext}} = \sigma_{\text{sca}} + \sigma_{\text{abs}}$) can be many times larger than the geometrical one, demonstrating the ability of plasmonic nanoparticles to focus far-field radiation into sub-wavelength volumes.

In realistic experimental conditions, however, photocatalytic nanoparticles are typically not perfectly spherical and are not embedded in a homogeneous medium, for example when a support is used. In fact, it has been shown that highly anisotropic particles with sharp tips, such as nanostars and octopods, show enhanced catalytic activity when compared to spheres [58–60]. To calculate the electric fields and the scattering and absorption cross sections for arbitrary geometries, Maxwell's equations need to be solved numerically, for example using a finite-difference time-domain method or a finite elements method.

The resonance condition $\epsilon = -2\epsilon_m$ shows that the spectral position of the LSPR can in principle be tuned by changing the permittivity of the surrounding medium (Figure 1a) or the composition of the particle (Figure 1c). In reality, however, these parameters cannot be set arbitrarily, as photocatalytic experiments are usually performed either in reactive gas environments or in aqueous solutions and the composition of the particle cannot be changed at will.

Nevertheless, for SERS the LSPR wavelength needs to be matched to the laser wavelength that is used to probe the molecular vibrations and for super-resolution fluorescence microscopy the LSPR wavelength needs to be spectrally separated from the emission of the reaction products to avoid mislocalization effects [61,62]. This necessary spectral tunability of the LSPR is typically achieved by synthesizing anisotropic particles such as rods, prisms, or cubes [63,64]. For example, for plasmonic nanorods an increasing aspect ratio results in a longer LSPR wavelength of their dominant resonance (Figure 1d).

The radiative decay of an LSPR (scattering) gives rise to intense electromagnetic fields on the nanoparticle surface, which are typically known as near-fields. The magnitude of these scattered fields can be amplified to values several orders of magnitude higher than the incoming electric field, depending on the nanoparticle morphology and composition. This phenomenon is responsible, for example, for the amplified Raman signal on plasmonic nanostructures, as will be discussed further in Section 3. The non-radiative decay of an LSPR (absorption) results in the formation of highly energetic non-equilibrium charge carriers which can drive redox reactions [2]. These hot electrons and holes are typically extremely short-lived, due to the high charge carrier density of metals [65,66]. When these carriers are not harvested, they dissipate their energy via electron-electron and electron-phonon scattering events [67], thereby heating up the particle and, eventually, its surrounding environment (Figure 1e) [2]. The latter photothermal process can also accelerate temperature-dependent chemical reactions according to the Arrhenius equation [68].

3. Surface-Enhanced Raman Spectroscopy

The radiative decay of plasmon resonances can result in a strong amplification of the electromagnetic fields at the surface of metal nanoparticles. Nanostructures possessing sharp corners and tips with low radii of curvature, or metal structures separated by nanoscale gaps such as dimers, can be utilized to amplify the magnitude of these plasmonic near-fields. For example, at resonant illumination, 75 nm Ag nanocube dimers separated by a 1 nm gap can sustain hot spots with electric field intensities up to 6 orders of magnitude higher than the incident radiation (Figure 2a) [69]. These electromagnetic hotspots can be exploited for a wide variety of applications ranging from driving photo-sensitive reactions [20,70–72], to enhancing efficiencies in photovoltaic devices [73], amplifying photoluminescence in semiconductors [74], and characterizing catalytic reactions occurring on the nanoparticle surfaces [53,75,76].

In particular, the Raman vibrational signals of molecules adsorbed on the nanoparticle can be strongly enhanced by the plasmonic near-fields, as these Raman signals are approximately proportional to the 4th power of the local electromagnetic field [77]. Such strong field dependence has its origin in the relatively broadband features of typical plasmon resonances (Figure 1c). The dipole-dipole interaction of the adsorbed molecule with the plasmonic near-fields, in fact, can enhance both light absorption at the Raman pump frequency as well as light emission at the slightly-shifted

Stokes and Anti-Stokes scattering frequencies [77,78]. Previously, near-field enhancements have been used to amplify the Raman signals of molecules up to 10^{10} times with respect to the signal in the absence of a plasmonic nanostructure [79]. This technique, referred to as surface-enhanced Raman spectroscopy (SERS), can provide improved spatial, temporal, and spectral resolution compared to normal Raman characterization of molecules [53,80–83]. Since the plasmonic near-fields are confined to a few nanometers from the nanoparticle surface, SERS selectively provides information of molecules adsorbed to the nanoparticle, hence minimizing background noise and pushing detection limits down to single molecules [84–88]. The large SERS signal enhancement has also enabled measurements with lower acquisition times, of the order of a few milliseconds, while maintaining a high signal-to-noise ratio, thereby allowing to monitor chemical and diffusion dynamics of surface adsorbates [84,89].

In the context of plasmon-driven catalysis, a single laser can be used to both drive catalytic reactions as well as characterize their SERS signals, thereby eliminating the need of complicated setups [90–92]. Below, we briefly describe the typical experimental setup of SERS measurements followed by a few examples of how this technique has been used to gain mechanistic understanding of plasmon-driven reactions.

Typically, single particle SERS measurements are performed by depositing dilute suspensions of colloidal plasmonic nanoparticles on transparent substrates to allow the interrogation of individual nanoparticles [46,53,93]. Alternatively, nanoparticles fabricated using a top-down approach such as focussed ion beam milling, nano-imprint lithography, electron-beam lithography, or hole-mask colloidal lithography can also be employed [45,94]. These nanoparticles are then illuminated with a focused laser beam, typically through an objective lens [46]. The laser photon energy is chosen to overlap with the LSPR, so as to obtain large near-field enhancements and thus high SERS signal-to-noise ratios.

The SERS detection of molecules on single metal nanoparticles can be performed in both transmission and reflection microscope configurations (Figure 2b). Since SERS sensitivity is the highest at the surface of the nanoparticle, it is necessary to properly identify its spatial position before irradiating it with a focused laser. As such, in both these geometries, first, a dark-field image of the substrate is typically obtained by focusing a broadband light source at large incident angles using an objective possessing a high numerical aperture (NA) [95]. The scattered light is then collected with an objective with a lower numerical aperture. Such an illumination geometry results in a dark background, where the plasmonic nanoparticles can be identified as small, diffraction-limited, bright scattering spots (Figure 2c). Once the nanoparticle position on the substrate is determined, a focused laser beam is used to excite the Raman vibrational modes of molecules adsorbed on its surface and the scattered light is guided to a CCD camera coupled with a grating. Often, a notch filter is kept in the scattering pathway, in order to exclude any laser damage to the spectrometer. The analyte of interest can either be adsorbed before depositing the nanoparticles on the substrate or it can be introduced in-situ using a flow cell configuration.

SERS measurements on single nanoparticles can provide valuable catalytic information such as the identification of reaction intermediates and the reaction kinetics [51,53,92,96]. Initial studies in this field focused on the model reduction of p-nitrobenzenethiol (NBT) to p-aminobenzenethiol (ABT) or p,p'-dimercaptoazobenzene (DMAB) catalyzed by metal nanoparticles. The advantage of employing such model reactions is that the reactants and products display strong Raman signals, which facilitate easy characterization of the process kinetics [45]. The absence of many competitive side reactions and the ability to operate at ambient conditions makes it a relatively easy system to study [97]. The strong affinity of the thiol functional group to the metal surface also ensure that the measured SERS signal originates exclusively from molecules adsorbed to the nanoparticles.

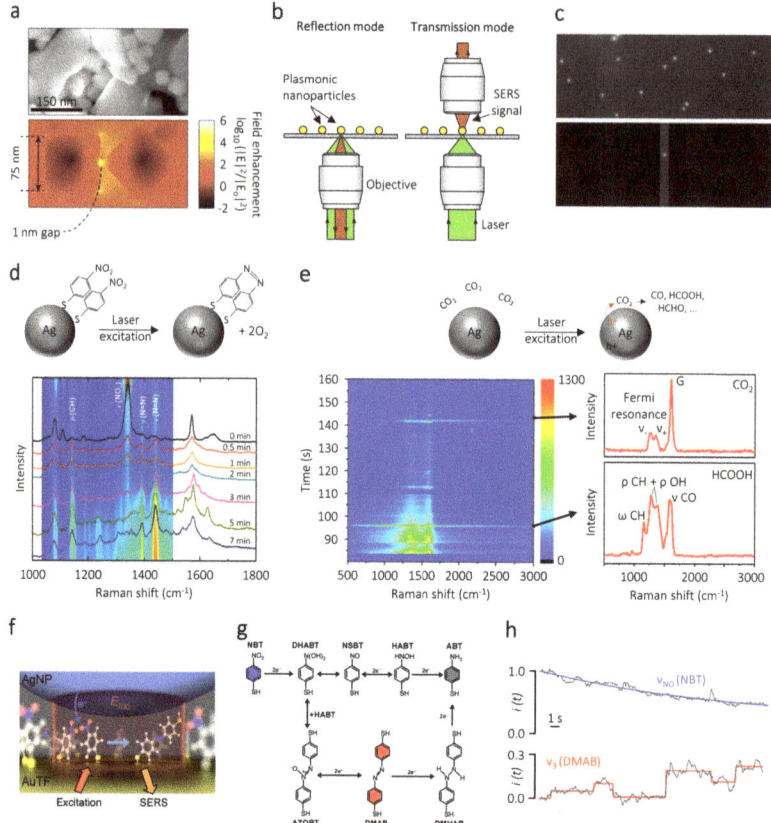

Figure 2. Surface-enhanced Raman spectroscopy (SERS). (a) (top) Scanning electron micrograph of silver nanocubes supported on alumina particles. The cubes arrange randomly, thereby generating geometries that can provide high electric fields. (bottom) Finite-difference time-domain simulation of the spatial distribution of the electric fields of 75 nm Ag nanocubes separated by 1 nm gap. The calculation is performed at the resonance wavelength of 500 nm. (b) Schematic illustration of SERS measurements over single nanoparticles in reflection and transmission mode. (c) (top) Grayscale dark-field image of several single plasmonic nanoparticles deposited on a transparent substrate. (bottom) Dark-field image of an individual nanoparticle obtained by closing the parallel slit located at the entrance of the monochromator, corresponding to the image plane outside the microscope. (d) (top) Schematic representation of the plasmon-driven reduction of p-nitrobenzenethiol (NBT) to p,p'-dimercaptoazobenzene (DMAB). (bottom) Time-resolved SERS measurement of the NBT reduction. (e) (top) Schematic representation of the plasmon-driven reduction of CO_2. (bottom) Time-resolved SERS measurement indicating the intermittent nature of the signals from the physisorbed CO_2 reactant and the catalytic product HCOOH. (f) Artistic representation of the nanoparticle on mirror geometry, wherein NBT molecules are sandwiched between a Ag nanoparticle and a Au thin film. (g) Direct and indirect reaction pathways for the reduction of NBT to ABT. (h) Time-resolved SERS intensities of NBT and DMAB, along with the fit to a single exponential function (blue) and step function (red). Panel (a) is reproduced with permission of [69]. Copyright Nature Publishing Group, 2012. Panel (d) is reproduced with permission of [80]. Copyright Royal Society of Chemistry, 2013. Panel (e) is reproduced with permission of [93]. Copyright American Chemical Society, 2018. Panels (f–h) are reproduced with permission of [84]. Copyright American Chemical Society, 2016.

The rate of these model reactions can be enhanced by plasmonic hot electrons. For example, Kang et al. employed single Ag microparticles (~2 μm diameter) with rough surfaces as SERS substrates to analyze the plasmon-driven reductive dimerization of NBT to DMAB (Figure 2d) [80]. Nanoparticles with rough surfaces have electromagnetic field enhancements that are significantly higher than their smooth counterparts and can therefore act as excellent Raman substrates. The plasmonic substrates were functionalized with NBT and then deposited on a silicon wafer, followed by laser illumination under atmospheric conditions. Time-resolved SERS spectra of the reduction reaction revealed the disappearance of the $\nu(NO_2,sym)$ peak of NBT (1335 cm^{-1}) as the reaction proceeded, along with an increase in the $\nu(N=N,sym)$ peak of DMAB (1440 cm^{-1}). To elucidate the influence of the illumination parameters on the reaction rate, SERS characterization was performed at various illumination powers using 532 nm and 633 nm continuous wave lasers. Under 532 nm illumination the reaction kinetics were significantly faster than under 633 nm illumination at constant laser power. This difference in the reactivity was attributed to the change in the absorbed power in the nanoparticle and the subsequent change in the number of hot charge carriers ejected. Strikingly, aminobenzenthiol (ABT) which has been reported as a reduction product of NBT was not observed in these experiments. To gain mechanistic insights in to the selectivity of the NBT reduction, the authors performed additional experiments under controlled environments. They observed that in the presence of H_2O or H_2, the DMAB products are further reduced into ABT [98]. Such experiments show how SERS can be used to follow reaction kinetics of catalytic reactions occuring on a single nanoparticle photocatalyst with controlled or predetermined geometry and surface sites.

Often, the exact plasmon-activation process behind photochemical reactions is challenging to identify, as the different competing mechanisms occur at ultra short timescales [22,28]. Photothermal effects have been used as an alternative explanation to many of the reported plasmon-driven chemical reactions [27]. SERS measurements offer an opportunity to quantitatively assess these thermal effects in both illuminated ensemble and single nanoparticle systems [99–101]. By comparing the Anti-Stokes (I_{AS}) and Stokes (I_S) intensity modes of the adsorbate, the vibrational temperature can be quantified using the relationship [101]:

$$\frac{I_{AS}}{I_S} = A \left[\frac{\tau \sigma'_s I}{h \nu_L} + \exp \frac{-h \nu_{vib}}{k_b T} \right] \quad (4)$$

where, A is a constant that takes into account the electric field enhancement at the nanoparticle surface and the Raman cross-section of the molecule, τ is the lifetime of the vibrational excited state, σ'_s is the Raman cross-section at the Stokes signal, I is the laser intensity, ν is the frequency, k_b is the Boltzmann constant, and T is the temperature experienced by the molecule. The subscripts L and vib represent laser and Raman vibrational mode respectively.

For example, Wu et al. ruled out the contribution of photothermal effects in the plasmon-driven electron transfer to [6,6]-phenyl-C_{61}-butyric acid methyl ester by extracting the temperature of their Au nanoparticle suspension using Equation (4) [100]. In this study, the temperature increase at various illumination intensities was found to be negligible, thereby confirming the electronic origin of plasmonic effects [100]. Similarly, Pozzi et al. investigated the temperature experienced by Rhodamine 6G molecules adsorbed onto single Ag nanoparticle aggregates using SERS measurements [101]. In their report, they highlighted that accurate temperature measurements using Equation (4) should be accompanied by careful estimation of the different electric field enhancements and molecular Raman cross-sections at the Stokes and Anti-Stokes frequencies, both of which contribute to the value of the constant A [101].

SERS measurements on single nanoparticles have also been used to provide mechanistic insights into plasmon-driven CO_2 reduction (Figure 2e) [93]. Kumari et al. chose Ag nanoparticles as SERS substrates for this purpose, as they typically display larger near-field enhancements compared to metals such as Au and Cu. In their experiments, a focused 514.5 nm laser was used to excite the SERS spectra of adsorbed reactant and product molecules on 60 nm Ag nanosphere aggregates dispersed on a glass substrate. Time-resolved SERS spectra displayed the generation of carbon monoxide, hydrogen, formic acid, formaldehyde, methanol, and carbonate molecules on the nanoparticle surface. The blinking nature of the SERS signals in this study suggests that only a single molecule is present over the nanoparticle surface at a time. Interestingly, the authors observed the generation of a surface-adsorbed hydrocarboxyl radical HOCO* during CO_2 reduction. Using density functional theory calculations, this intermediate was found to be crucial in determining the selectivity of the photoreduced products. Thanks to the extreme field confinement on plasmonic nanoparticles, single particle studies can detect the stochastic adsorption and desorption of reactants and products along with the formation of intermediate species at millisecond time resolutions. Such single molecule and real-time detection capability is unachievable using ensemble studies due to signal averaging. Single particle measurements focused on identifying the reaction intermediates will play a pivotal role in elucidating structure-activity relationships of future nanostructured photocatalysts.

Recently, SERS measurements have also been employed to obtain mechanistic insights into plasmon-driven ethylene epoxidation on Ag nanospheres [36]. The time evolution of the Raman signals indicated a correlation between the formation of a graphene layer on the nanoparticle surface and the formation of ethylene oxidation species. Further experiments using a graphene coated Ag nanoparticle, along with DFT calculations, confirmed that the in-situ formed graphene, rather than the Ag nanoparticle, is the catalyst for ethylene epoxidation. Such mechanistic insight could not have been obtained with ensemble measurements, in which it would have been impossible to distinguish the narrow signature of a graphene nanocrystal among the broad Raman spectra of mixed carbonaceous species typically obtained upon ensemble averaging. Understanding the role of graphene nanocrystals in ethylene epoxidation could inform the development of industrial photocatalysts, as ethylene oxide is an important feedstock in the synthesis of solvents, plastics, and other organic chemicals.

Choi et al. conclusively demonstrated single molecule plasmon-driven conversion of NBT to ABT, by sandwiching the reactant molecules between a Au thin film and a Ag NP (Figure 2f) [84]. Such geometries are commonly referred to as nanoparticle on mirror (NPoM) or as particle over surface. The nanoscale gap between the nanospheres and the film generates a SERS enhancement factor of 10^8 that varies by less than an order of magnitude between individual particles. Such reproducibility in the SERS geometries can be attributed to the fixed length of the spacer molecule, in this case NBT. The large signal-to-noise ratio of the SERS spectra allowed the authors to perform measurements at millisecond time scales, thereby accurately quantifying the reaction kinetics. In this study, the SERS measurements displayed discrete jumps in the DMAB intermediate signal while the NBT reactant signal continuously decayed. The discrete spectral jumps indicate the conversion of a single molecule, which was further corroborated with a kinetic Monte Carlo model accounting for the spatial variations in the near-field enhancements and the Raman cross sections of the molecules.

The authors also used the kinetic information to obtain mechanistic insights into the reduction of NBT. Two main reaction mechanisms had been previously proposed for the conversion of NBT to ABT, namely direct and indirect (Figure 2g). The direct mechanism generated ABT via the formation of dihydroxyaminobenzenethiol (DHABT), nitrosobenzenethiol (NSBT), and hydroxylaminobenzenethiol (HABT) intermediates, while the indirect path involved condensation of DHABT and HABT to form DMAB and its subsequent reduction to ABT. By analysing the rate of NBT consumption and the rate of DMAB formation, the authors pointed out that only less than 10% of NBT molecules undergo reduction via the indirect path (Figure 2h). Such deep insights into the reaction mechanism reveals the power of SERS measurements to identify spectro-temporal behavior in single molecule photocatalytic experiments, without the addition of any label molecules.

4. Super-Resolution Fluorescence Microscopy

Optical fluorescence microscopy is a powerful imaging technique typically used for the structural characterization of biological samples. However, due to diffraction its spatial resolution remains limited to $\sim \lambda/2NA$ [102], corresponding to $\sim$200 nm for visible wavelengths λ and objectives with high numerical aperture (NA $\approx$ 1.4). This limitation in resolution can be overcome when imaging the fluorescence of a single molecule. Under the condition of a single fluorophore, in fact, the measured emission pattern can be fit to a two-dimensional Gaussian, where the center of the Gaussian is the position of the molecule [103]. The localization precision σ_{loc} of this position can be calculated using [104]:

$$\sigma_{\text{loc}}^2 = \frac{\sigma_a^2}{N}\left(\frac{16}{9} + \frac{8\pi\sigma_a^2 b^2}{Na^2}\right), \tag{5}$$

$$\sigma_a^2 = \sigma^2 + a^2/12, \tag{6}$$

where σ is the standard deviation of the Gaussian, a is the size of a camera pixel, N is the number of detected photons, and b is the standard deviation in the background noise. Under the experimental conditions discussed in this Perspective, σ_{loc} is typically between 10 and 30 nm, which is commensurate with the typical size of plasmonic photocatalysts.

Although originally developed to image sub-diffraction limited features in biological samples [105–107], the use of super-resolution microscopy has recently been extended to a variety of research fields, from the optimization of nanophotonic devices [108–110], to the characterization of nanomaterials [111]. Recently, single molecule localization has been extended to the mapping of heterogeneous chemical reactions with nanoscale spatial resolution and single molecule turnover accuracy. For example, super-resolution microscopy has been used to study so-called fluorogenic reactions, catalytic conversions producing a fluorescent reaction product [112]. In a typical experiment, catalysts are dropcasted on a glass substrate, which is built into a flowcell that allows a continuous supply of reactants. The catalysts are dropcasted at a sufficiently low concentration, such that individual particles can easily be resolved spatially. The sample is then illuminated with a laser that excites the fluorescence of the reaction products. To minimize background signal, this illumination is usually done in a total internal reflection fluorescence (TIRF) configuration [113], either using the same high NA objective that is used to capture the fluorescence (Figure 3a), or using a prism (Figure 3b).

Illumination through the objective (Figure 3a) requires the NA to be higher than the critical angle for TIRF, which implies NA > 1.333, assuming the surrounding medium is water. In practice, an NA of ∼1.45 is typically used, as otherwise only a very small part of the peripheral area of the lens can be used for TIRF illumination, making the alignment procedure challenging. Despite being more costly, an objective with a higher NA also results in a superior collection efficiency thanks to its high acceptance angle. Since the localization precision scales with $1/\sqrt{N}$ (see Equation (5)), a higher collection efficiency improves the localization precision. Furthermore, illumination through the objective allows for easy access to the sample from the top. Illumination through a prism (Figure 3b) has a lower collection efficiency due to the typically lower NA objectives used and can suffer from light scattering by impurities in the liquid. On the other hand, this illumination method is less costly, as it does not require extremely high NA, and suffers less from fluorescence quenching, as the nanoparticle catalyst is not in the detection pathway.

Both illumination methods are widefield, which allows many particles to be measured simultaneously (Figure 3c). Since the typical integration time that is needed to acquire enough photons to localize a single fluorophore is between 10 and 100 ms, only fluorescent molecules that are adsorbed on the catalyst are detected. Fluorescent species that are free in solution, in fact, move too fast due to Brownian motion and only contribute to the background signal. In fluorogenic reactions, upon product formation a fluorescent burst is detected, until the product molecule desorbs from the catalyst surface and diffuses away (Figure 3d). Each fluorescent burst (Figure 3e) is fitted to a two-dimensional Gaussian (Figure 3f), which gives the in-plane position of each molecule with nanometer accuracy.

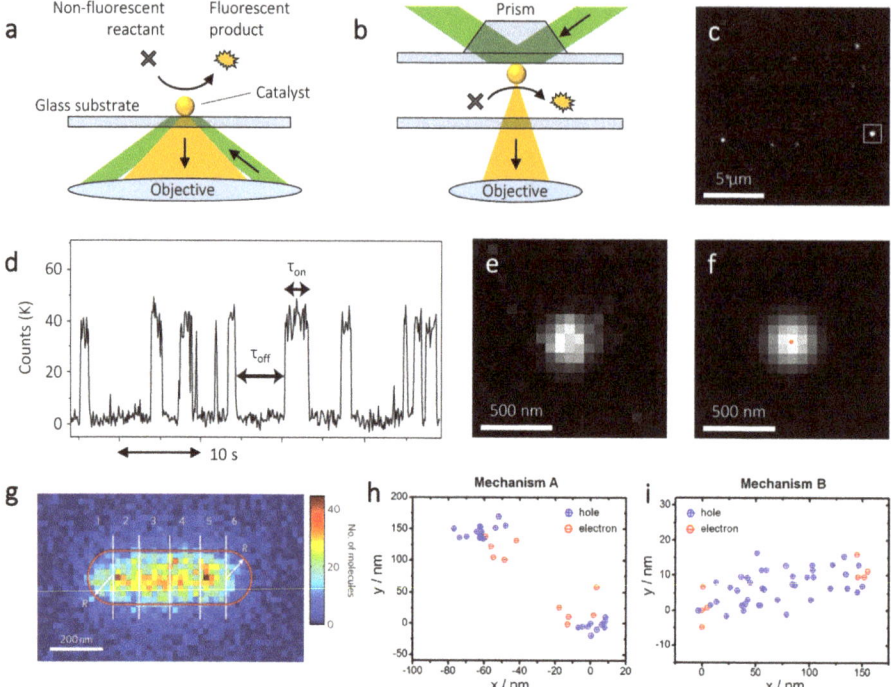

Figure 3. Super-resolution catalysis mapping on individual metal nanoparticles. (**a,b**) A non-fluorescent reactant converts to a fluorescent product, which can be imaged in an optical fluorescence microscope. Total internal reflection fluorescence (TIRF) illumination is achieved through the objective (**a**) or through a prism (**b**). (**c**) Widefield image with fluorescent events on multiple particles. (**d**) Time trace of the fluorescence intensity measured on a single catalytic particle; τ_{off} is the product formation time and τ_{on} is the product desorption time. (**e,f**) A single fluorescent burst (**e**), corresponding to a single reaction product, can be fit to a two-dimensional Gaussian (**f**), which results in the position of the molecule (red dot in panel f). (**g**) Two-dimensional histogram of resorufin molecules detected on a single gold nanorod. The nanorod is divided in 6 segments and for each segment a specific turnover rate can be calculated. (**h,i**) Super-resolved locations of resorufin molecules detected on a single CdS-Au nanorod. Blue points correspond to hole-driven reactions and red points to electron-driven reaction. The nanorod is excited below bandgap with a 532 nm laser (**h**) or above bandgap with a 405 nm laser (**i**). Panels (**c,e,f**) are reproduced with permission of [110]. Copyright American Chemical Society, 2019. Panel (**d**) is reproduced with permission of [29]. Copyright Nature Publishing Group, 2008. Panel (**g**) is reproduced with permission of [31]. Copyright Nature Publishing Group, 2012. Panels (**h,i**) are reproduced with permission of [114]. Copyright American Chemical Society, 2014.

In 2008, Xu et al. reported the real-time imaging of redox catalysis on ∼6 nm gold particles with a time resolution up to 30 ms [29]. The reaction under study was the gold-catalyzed reduction of the nonfluorescent molecule resazurin to the highly fluorescent molecule resorufin in the presence of NH_2OH as a reducing agent. Despite the fairly narrow size distribution of the gold particles (6.0 ± 1.7 nm), the study revealed a large heterogeneity in turnover rate. Furthermore, the authors showed that within a single batch of nanoparticles different reaction pathways exist for the desorption of the resorufin molecules from the gold surface. While most particles preferred product desorption assisted by a reactant binding step, some preferred a direct desorption of the product molecule resorufin, and some others showed no preference between the two mechanisms. This diverse behavior demonstrates the heterogeneity in catalytic properties of metal nanoparticles and highlights the importance of single particle approaches to unveil their structure-function relationship. This study was later extended to particles of varying sizes up to ~14 nm, which showed that larger particles have a lower turnover rate when normalized to the surface area, are less selective between the different desorption mechanisms, and are less prone to restructuring [115].

In both of these studies the size of the catalyst was smaller than the spatial resolution typically achievable with super-resolution microscopy, which hinders the imaging of spatial heterogeneities within a single catalytic nanoparticle. For larger catalysts, however, their activity can be resolved spatially. Figure 3g shows a two-dimensional histogram of the detected resorufin molecules catalytically produced at the surface of a single Au nanorod [31]. By segmenting individual nanorods and determining the specific turnover rate for each segment, it was found that the nanorod tips generally have higher reactivity than their cores. This difference in reactivity can be attributed to the presence of more corner and edge sites at the nanorod tips, which are typically more reactive. Interestingly, when only analyzing the reactivity of the nanorod core, i.e., without the tips, the authors found a specific turnover rate that decreased with increasing distance from the center, even though the core is made up of the same facets. This reactivity gradient can be attributed to the growth mechanism of the nanorod. During synthesis, the growth rate of the nanorod decreases linearly with increasing length [116]. Since faster colloidal growth is typically accompanied by a higher density of superficial defects, a gradient in defect density develops, resulting in higher reactivity at the nanorod center [31]. These sub-particle catalysis maps demonstrate that, besides the types of surface facets [30], defects play a large role in the catalytic activity of metal nanoparticles [31,32].

Additionally, sub-particle spatial resolution allows the visualization of spatiotemporal variations in catalytic activity due to surface restructuring. Measurements of the catalytic activity of triangular Au nanoplates showed that the spatial distribution of catalytic events slowly varies over timescales of several hours [34]. Although the reactivity between different corners of the same nanoplate was found to initially be heterogeneous, over time all corners converged to a similar reactivity. This slow loss of heterogeneity was attributed to the dynamic restructuring of the nanoplate corners from sharp to blunt, as also confirmed with ex-situ TEM [34].

Lastly, spatiotemporally-resolved single molecule catalysis imaging has recently been used to study the correlation between events on a single catalyst, and between events on different catalysts [117]. The authors found that catalytic reactions on individual Pd or Au nanoparticles were correlated over lengths scales of ~100 nm, probably via the transport of positively charged holes over the catalyst surface. The events between different nanoparticles was also found to be correlated over many micrometers via the molecular diffusion of negatively charged reaction products. Both the influence of defects and of dynamic restructuring on catalytic activity and the correlation between different events would have remained hidden in ensemble or diffraction-limited measurements.

Although the fluorogenic catalytic reactions used in these seminal studies typically also happen in the dark, their rates can be enhanced under laser excitation [114,118,119]. Measurements on CdS semiconductor nanorods decorated with Au nanoparticle co-catalysts at their tips showed that plasmon excitation (below bandgap) and semiconductor excitation (above bandgap) both enhance the rate of Amplex Red oxidation to resorufin [114]. Photogenerated electrons promote desorption of the negatively charged resorufin molecule due electrostatic force. Therefore, electron-driven turnovers are characterized by short fluorescent bursts. Conversely, photogenerated holes enhance adsorption of resorufin and, therefore, increase the length of the fluorescent bursts. On the CdS-Au nanostructures two distinct burst lengths were observed, which allowed the authors to distinguish between electron-driven and hole-driven reactions. When the plasmon resonance of the Au nanoparticles is excited (below bandgap) hot electrons and holes are generated within the Au. The photogenerated electron is injected into the conduction band of the semiconductor due to the Schottky barrier that is formed at the CdS-Au interface, and the photogenerated hole remains in the Au. This charge separation is also observed in the super-resolution catalysis maps (Figure 3h), with the hole-driven reactions (long bursts) taking place on the edges of the rod where the Au nanoparticles are located, and the electron-driven reactions (short bursts) taking place a few tens of nanometers inward on the CdS nanorod. For above bandgap excitation, charge carriers are generated in the CdS nanorod. The photogenerated electrons transfer to the Au nanoparticles, whereas the holes remain in the nanorod. This charge separation is again reflected in the super-resolution catalysis maps (Figure 3h), with the hole-driven reactions now taking place all over the nanorod and the electron-driven reactions on the Au nanoparticles [114].

On dimers of Au nanorods with nanoscale gaps, the turnover rate of resazurin to resorufin reduction was also found to be dependent on the incident laser power [118]. The turnover rate scaled quadratically with laser power and, therefore, the enhancement can be attributed to the presence of two photoexcited species on the nanoparticle surface. However, the incident laser photoexcites both the reactant resazurin and the plasmon resonance of the catalysts, preventing a straightforward distinction between near-field effects (photoexcitation of the reactants) and hot charge carrier effects [118].

Another approach of characterizing the influence of plasmon excitation on the catalytic reactivity of nanoparticles is by measuring the change in the single particle activation barrier of the fluorogenic reaction when the plasmon resonance is excited. According to the Arrhenius equation, both the product formation and the product desorption rates, $\langle \tau_{off} \rangle^{-1}$ and $\langle \tau_{on} \rangle^{-1}$, respectively, depend on temperature via [120]

$$\langle \tau_{off} \rangle^{-1} = A_{off} \exp[-E_{a,off}/RT], \tag{7}$$

$$\langle \tau_{on} \rangle^{-1} = A_{on} \exp[-E_{a,on}/RT], \tag{8}$$

where A_{off} and A_{on} are prefactors, $E_{a,off}$ and $E_{a,on}$ are the activation barriers for the product formation and desorption processes, respectively, R is the gas constant, and T is the temperature. Therefore, by measuring $\langle \tau_{off} \rangle$ and $\langle \tau_{on} \rangle$ for varying reactor temperatures, the single particle activation barriers for the product formation and desorption can be extracted [120,121]. For chemical reactions with an intermediate species, also a distinction can be made between the activation barrier for intermediate formation and for final product formation [119]. Single molecule experiments on the Au nanorod catalyzed oxidation of Amplex Red to resorufin showed that upon plasmon excitation, the activation barrier for intermediate formation decreased, while the activation barriers for product formation and product desorption remained unaltered [119]. This lowering in activation barrier was attributed to the presence of hot charge carriers. Thermal effects were ruled out based on calculations and thermocouple measurements. Furthermore, no change in activation barrier was observed for the resazurin reduction to resorufin, which would be expected if the plasmonic enhancement was purely thermal [119].

One obvious limitation of the super-resolution fluorescence microscopy studies presented so far is that the reaction under study needs to generate a fluorescent molecule. Most biological and chemical processes, however, do not involve fluorescent species. This limitation has recently been overcome using a competition-based technique, in which a single nanoparticle can catalyze two chemical conversions [122]. The first reaction is the reaction of interest and its reactants and products do not fluoresce. The second auxiliary reaction is fluorogenic and can be imaged and localized with nanometer spatial resolution. If both reactions compete for the same surface sites, the reaction of interest suppresses the rate of the fluorogenic reaction. The extent of suppression can be imaged using super-resolution microscopy, thereby giving spatial information on the reaction of interest.

5. Outlook

We began this Perspective by introducing the optical properties of metallic nanoparticles. We discussed how the decay of plasmon resonances generates hot charge carriers, near-field enhancements, and elevated nanoparticle surface temperatures, all of which can simultaneously contribute to drive chemical reactions. We have then introduced two far-field optical microscopy techniques, surface-enhanced Raman spectroscopy and super-resolution fluorescence microscopy, and we have highlighted how they can disentangle the activation mechanism of plasmon-driven chemical reactions thanks to their unique temporal and spatial resolutions. In this final outlook section, we provide our perspective on how these techniques can be further used to tackle current open questions in the field of plasmon-driven chemistry.

As we have seen, the resazurin-to-resorufin conversion is often used to benchmark plasmonic photocatalysts at the single particle level. Super-resolution fluorescence microscopy measurements suggest that hot charge carriers are the main driving force behind the increased turnover rate observed under plasmon excitation. However, the reaction product resorufin and the plasmon resonance of the metal nanoparticle are typically excited using the same 532 nm laser. At this wavelength the reactant resazurin is weakly absorbing, which can lead to its photodriven disproportionation to resorufin. This side reaction is enhanced by the strong plasmonic near-fields at the nanoparticle surface and can therefore entirely by-pass any photocatalytic charge transfer process [123]. To decouple plasmon-driven charge transfer effects from the photoexcitation of reactant molecules, it would be desirable to design experiments involving spectrally detuned plasmonic catalysts and two lasers, one for the excitation of the fluorescent molecules and one for the plasmon resonance of the catalyst [118,119].

Furthermore, to confidently eliminate the contribution of plasmonic heating in both SERS and super-resolution measurements, control experiments should be performed where the nanoparticle system is externally heated to the measured temperature [27]. Ideally, in-situ experimental characterization of the temperature should be performed [101,124,125], for example by analyzing the Anti-Stokes and Stokes signals of the SERS spectra. In the absence of experimental measurements, theoretical calculations can give an estimate of the surface temperature of the irradiated nanoparticle [125–128]. Although such experiments may seem obvious, they are often challenging in practice, especially when studying single nanoparticles using an optical microscope. Under these conditions, in fact, laser irradiation can easily heat the nanoparticle under study to hundreds of degrees Celsius and these elevated temperatures are difficult to reach with conventional thermally-controlled sample holders. The lack of simple thermometric techniques at the nanoscale, along with the challenges in accurately calculating temperature increases under *operando* conditions, make it difficult to effectively assess the temperature at the active site of a plasmonic photocatalyst [127,129].

To establish structure-activity relationships, it is crucial to characterize the morphology of the nanocatalysts. This can be done using dark-field scattering spectroscopy or electron microscopy before, during, and after the catalytic reaction, along with the SERS or super-resolution characterization [38]. Most SERS studies on catalytic reactions are performed on nanoparticle aggregates rather than single particles. Extending SERS measurements to single particles could allow detecting catalytic intermediates with higher signal-to-noise ratios, and establishing more accurate structure-activity relationships. Another strategy to ensure that these measurements are performed on individual structures with a controlled size is by utilizing nanoparticle substrates fabricated using a top-down approach such as electron-beam lithography [45,130–132]. An additional advantage of using top-down techniques to deposit plasmonic photocatalysts is that the surface of the nanoparticles is typically free of contaminants or ligands, hence minimizing their potential contribution to the measured catalytic activity.

Lastly, plasmonic enhancements are usually characterized by the single particle turnover rate, given by the number of detected products per second. However, the nanoscale spatial distribution of catalytic events, as unveiled by super-resolution fluorescence microscopy, can also provide valuable information on the underlying activation mechanism [129,133]. Both the plasmonic near-fields and the generation of non-equilibrium charge carriers, in fact, are strongly polarization dependent [67,134], whereas the nanoparticle temperature is expected to be homogeneous due to the high thermal conductivity of the metal [2,133,135].

Studying light-driven reactions on metal photocatalysts in real-time and with single particle spatial resolution is a challenging task that promises to unveil previously inaccessible information on the photocatalyst active sites, the photoactivation mechanism, the molecular intermediates, and the reaction pathways. Here we have presented recent successes and future promises of two far-field optical techniques, that have recently allowed us a closer look on these fundamental nanoscale processes.

Author Contributions: R.F.H. and R.K. wrote the first draft of the manuscript. All authors edited the manuscript. All authors have read and agreed to the published version of the manuscript.

Funding: This work was supported by the Dutch Research Council (Nederlandse Organisatie voor Wetenschappelijk Onderzoek) via the NWO Vidi award 680-47-550.

Acknowledgments: The authors acknowledge Anne Askey for assisting in writing the section on Surface-enhanced Raman spectroscopy and Ferry Nugroho for giving feedback on the manuscript.

Conflicts of Interest: The authors declare no conflict of interest.

Abbreviations

The following abbreviations are used in this manuscript:

ABT	Aminobenzethiol
DHABT	Dihydroxyaminobenzenethiol
DMAB	p,p'-dimercaptoazobenzene
EF	Enhancement factor
HABT	Hydroxylaminobenzenethiol
LSPR	Localized surface plasmon resonance
NA	Numerical aperature
NPoM	Nanoparticle on mirror
NSBT	Nitrosobenzenethiol
NBT	p-nitrobenzenethiol
SERS	Surface-enhanced Raman spectroscopy
TEM	Transmission electron microscopy
TIRF	Total internal reflection fluorescence

References

1. Linic, S.; Christopher, P.; Ingram, D.B. Plasmonic-metal Nanostructures for Efficient Conversion of Solar to Chemical Energy. *Nat. Mater.* **2011**, *10*, 911–921. [CrossRef] [PubMed]
2. Brongersma, M.L.; Halas, N.J.; Nordlander, P. Plasmon-induced Hot Carrier Science and Technology. *Nat. Nanotechnol.* **2015**, *10*, 25. [CrossRef] [PubMed]
3. Linic, S.; Aslam, U.; Boerigter, C.; Morabito, M. Photochemical Transformations on Plasmonic Metal Nanoparticles. *Nat. Mater.* **2015**, *14*, 567–576. [CrossRef] [PubMed]
4. Zhang, Y.; He, S.; Guo, W.; Hu, Y.; Huang, J.; Mulcahy, J.R.; Wei, W.D. Surface-plasmon-driven Hot Electron Photochemistry. *Chem. Rev.* **2017**, *118*, 2927–2954. [CrossRef] [PubMed]
5. Aslam, U.; Rao, V.G.; Chavez, S.; Linic, S. Catalytic Conversion of Solar to Chemical Energy on Plasmonic Metal Nanostructures. *Nat. Catal.* **2018**, *1*, 656–665. [CrossRef]
6. Zhang, Z.; Zhang, C.; Zheng, H.; Xu, H. Plasmon-driven Catalysis on Molecules and Nanomaterials. *Acc. Chem. Res.* **2019**, *52*, 2506–2515. [CrossRef] [PubMed]
7. Thomann, I.; Pinaud, B.A.; Chen, Z.; Clemens, B.M.; Jaramillo, T.F.; Brongersma, M.L. Plasmon Enhanced Solar-to-fuel Energy Conversion. *Nano Lett.* **2011**, *11*, 3440–3446. [CrossRef] [PubMed]
8. Thimsen, E.; Le Formal, F.; Gratzel, M.; Warren, S.C. Influence of Plasmonic Au Nanoparticles on the Photoactivity of Fe2o3 Electrodes for Water Splitting. *Nano Lett.* **2011**, *11*, 35–43. [CrossRef]
9. Mubeen, S.; Lee, J.; Singh, N.; Krämer, S.; Stucky, G.D.; Moskovits, M. An Autonomous Photosynthetic Device in Which All Charge Carriers Derive from Surface Plasmons. *Nat. Nanotechnol.* **2013**, *8*, 247–251. [CrossRef]
10. Knight, M.W.; Sobhani, H.; Nordlander, P.; Halas, N.J. Photodetection with Active Optical Antennas. *Science* **2011**, *332*, 702–704. [CrossRef]
11. Liu, Y.; Cheng, R.; Liao, L.; Zhou, H.; Bai, J.; Liu, G.; Liu, L.; Huang, Y.; Duan, X. Plasmon Resonance Enhanced Multicolour Photodetection by Graphene. *Nat. Commun.* **2011**, *2*, 1–7. [CrossRef] [PubMed]

12. Huang, J.A.; Luo, L.B. Low-Dimensional Plasmonic Photodetectors: Recent Progress and Future Opportunities. *Adv. Opt. Mater.* **2018**, *6*, 1701282. [CrossRef]
13. Marimuthu, A.; Zhang, J.; Linic, S. Tuning Selectivity in Propylene Epoxidation by Plasmon Mediated Photo-switching of Cu Oxidation State. *Science* **2013**, *339*, 1590–1593. [CrossRef] [PubMed]
14. Zhang, X.; Li, X.; Zhang, D.; Su, N.Q.; Yang, W.; Everitt, H.O.; Liu, J. Product Selectivity in Plasmonic Photocatalysis for Carbon Dioxide Hydrogenation. *Nat. Commun.* **2017**, *8*, 1–9. [CrossRef]
15. Yu, S.; Wilson, A.J.; Kumari, G.; Zhang, X.; Jain, P.K. Opportunities and Challenges of Solar-energy-driven Carbon Dioxide to Fuel Conversion with Plasmonic Catalysts. *ACS Energy Lett.* **2017**, *2*, 2058–2070. [CrossRef]
16. DuChene, J.S.; Tagliabue, G.; Welch, A.J.; Li, X.; Cheng, W.H.; Atwater, H.A. Optical Excitation of a Nanoparticle Cu/p-NiO Photocathode Improves Reaction Selectivity for CO_2 Reduction in Aqueous Electrolytes. *Nano Lett.* **2020**, *20*, 2348–2358. [CrossRef]
17. Mukherjee, S.; Libisch, F.; Large, N.; Neumann, O.; Brown, L.V.; Cheng, J.; Lassiter, J.B.; Carter, E.A.; Nordlander, P.; Halas, N.J. Hot Electrons Do the Impossible: Plasmon-induced Dissociation of H_2 on Au. *Nano Lett.* **2013**, *13*, 240–247. [CrossRef]
18. Zhou, L.; Swearer, D.F.; Zhang, C.; Robatjazi, H.; Zhao, H.; Henderson, L.; Dong, L.; Christopher, P.; Carter, E.A.; Nordlander, P.; et al. Quantifying Hot Carrier and Thermal Contributions in Plasmonic Photocatalysis. *Science* **2018**, *362*, 69–72. [CrossRef]
19. Wu, K.; Chen, J.; McBride, J.R.; Lian, T. Efficient Hot-electron Transfer by a Plasmon-induced Interfacial Charge-transfer Transition. *Science* **2015**, *349*, 632–635. [CrossRef]
20. Boerigter, C.; Campana, R.; Morabito, M.; Linic, S. Evidence and Implications of Direct Charge Excitation as the Dominant Mechanism in Plasmon-mediated Photocatalysis. *Nat. Commun.* **2016**, *7*, 1–9. [CrossRef]
21. Boerigter, C.; Aslam, U.; Linic, S. Mechanism of Charge Transfer from Plasmonic Nanostructures to Chemically Attached Materials. *ACS Nano* **2016**, *10*, 6108–6115. [CrossRef] [PubMed]
22. Kamarudheen, R.; Castellanos, G.W.; Kamp, L.P.J.; Clercx, H.J.H.; Baldi, A. Quantifying Photothermal and Hot Charge Carrier Effects in Plasmon-driven Nanoparticle Syntheses. *ACS Nano* **2018**, *12*, 8447–8455. [CrossRef] [PubMed]
23. Yu, Y.; Sundaresan, V.; Willets, K.A. Hot Carriers Versus Thermal Effects: Resolving the Enhancement Mechanisms for Plasmon-mediated Photoelectrochemical Reactions. *J. Phys. Chem. C* **2018**, *122*, 5040–5048. [CrossRef]
24. Zhang, X.; Li, X.; Reish, M.E.; Zhang, D.; Su, N.Q.; Gutieérrez, Y.; Moreno, F.; Yang, W.; Everitt, H.O.; Liu, J. Plasmon-enhanced Catalysis: Distinguishing Thermal and Nonthermal Effects. *Nano Lett.* **2018**, *18*, 1714–1723. [CrossRef]
25. Aizpurua, J.; Ashfold, M.; Baletto, F.; Baumberg, J.; Christopher, P.; Cortés, E.; de Nijs, B.; Fernandez, Y.D.; Gargiulo, J.; Gawinkowski, S.; et al. Dynamics of Hot Electron Generation in Metallic Nanostructures: General Discussion. *Faraday Discuss.* **2019**, *214*, 123–146. [CrossRef]
26. Jain, P.K. Taking the Heat Off of Plasmonic Chemistry. *J. Phys. Chem. C* **2019**, *123*, 24347–24351. [CrossRef]
27. Sivan, Y.; Baraban, J.; Un, I.W.; Dubi, Y. Comment on "Quantifying Hot Carrier and Thermal Contributions in Plasmonic Photocatalysis". *Science* **2019**, *364*, eaaw9367. [CrossRef]
28. Kamarudheen, R.; Aalbers, G.J.W.; Hamans, R.F.; Kamp, L.P.J.; Baldi, A. Distinguishing Among All Possible Activation Mechanisms of a Plasmon-Driven Chemical Reaction. *ACS Energy Lett.* **2020**, *5*, 2605–2613. [CrossRef]
29. Xu, W.; Kong, J.S.; Yeh, Y.T.E.; Chen, P. Single-molecule Nanocatalysis Reveals Heterogeneous Reaction Pathways and Catalytic Dynamics. *Nat. Mater.* **2008**, *7*, 992–996. [CrossRef]
30. Chen, T.; Chen, S.; Song, P.; Zhang, Y.; Su, H.; Xu, W.; Zeng, J. Single-molecule Nanocatalysis Reveals Facet-dependent Catalytic Kinetics and Dynamics of Pallidium Nanoparticles. *ACS Catal.* **2017**, *7*, 2967–2972. [CrossRef]
31. Zhou, X.; Andoy, N.M.; Liu, G.; Choudhary, E.; Han, K.S.; Shen, H.; Chen, P. Quantitative Super-resolution Imaging Uncovers Reactivity Patterns on Single Nanocatalysts. *Nat. Nanotechnol.* **2012**, *7*, 237. [CrossRef] [PubMed]
32. Andoy, N.M.; Zhou, X.; Choudhary, E.; Shen, H.; Liu, G.; Chen, P. Single-molecule Catalysis Mapping Quantifies Site-specific Activity and Uncovers Radial Activity Gradient on Single 2D Nanocrystals. *J. Am. Chem. Soc.* **2013**, *135*, 1845–1852. [CrossRef] [PubMed]

33. Hansen, P.L.; Wagner, J.B.; Helveg, S.; Rostrup-Nielsen, J.R.; Clausen, B.S.; Topsøe, H. Atom-resolved Imaging of Dynamic Shape Changes in Supported Copper Nanocrystals. *Science* **2002**, *295*, 2053–2055. [CrossRef] [PubMed]
34. Zhang, Y.; Lucas, J.M.; Song, P.; Beberwyck, B.; Fu, Q.; Xu, W.; Alivisatos, A.P. Superresolution Fluorescence Mapping of Single-nanoparticle Catalysts Reveals Spatiotemporal Variations in Surface Reactivity. *Proc. Natl. Acad. Sci. USA* **2015**, *112*, 8959–8964. [CrossRef]
35. Zugic, B.; Wang, L.; Heine, C.; Zakharov, D.N.; Lechner, B.A.; Stach, E.A.; Biener, J.; Salmeron, M.; Madix, R.J.; Friend, C.M. Dynamic Restructuring Drives Catalytic Activity on Nanoporous Gold–silver Alloy Catalysts. *Nat. Mater.* **2017**, *16*, 558–564. [CrossRef]
36. Zhang, X.; Kumari, G.; Heo, J.; Jain, P.K. In Situ Formation of Catalytically Active Graphene in Ethylene Photo-epoxidation. *Nat. Commun.* **2018**, *9*, 1–10. [CrossRef]
37. Buurmans, I.L.C.; Weckhuysen, B.M. Heterogeneities of Individual Catalyst Particles in Space and Time as Monitored by Spectroscopy. *Nat. Chem.* **2012**, *4*, 873. [CrossRef]
38. Novo, C.; Funston, A.M.; Mulvaney, P. Direct Observation of Chemical Reactions on Single Gold Nanocrystals Using Surface Plasmon Spectroscopy. *Nat. Nanotechnol.* **2008**, *3*, 598–602. [CrossRef]
39. Urakawa, A.; Baiker, A. Space-resolved Profiling Relevant in Heterogeneous Catalysis. *Top. Catal.* **2009**, *52*, 1312–1322. [CrossRef]
40. Albinsson, D.; Bartling, S.; Nilsson, S.; Ström, H.; Fritzsche, J.; Langhammer, C. Operando Detection of Single Nanoparticle Activity Dynamics Inside a Model Pore Catalyst Material. *Sci. Adv.* **2020**, *6*, eaba7678. [CrossRef]
41. Beale, A.M.; Jacques, S.D.; Weckhuysen, B.M. Chemical Imaging of Catalytic Solids with Synchrotron Radiation. *Chem. Soc. Rev.* **2010**, *39*, 4656–4672. [CrossRef] [PubMed]
42. Amenabar, I.; Poly, S.; Nuansing, W.; Hubrich, E.H.; Govyadinov, A.A.; Huth, F.; Krutokhvostov, R.; Zhang, L.; Knez, M.; Heberle, J.; et al. Structural Analysis and Mapping of Individual Protein Complexes by Infrared Nanospectroscopy. *Nat. Commun.* **2013**, *4*, 1–9. [CrossRef] [PubMed]
43. Berweger, S.; Nguyen, D.M.; Muller, E.A.; Bechtel, H.A.; Perkins, T.T.; Raschke, M.B. Nano-chemical Infrared Imaging of Membrane Proteins in Lipid Bilayers. *J. Am. Chem. Soc.* **2013**, *135*, 18292–18295. [CrossRef] [PubMed]
44. Zhang, R.; Zhang, Y.; Dong, Z.; Jiang, S.; Zhang, C.; Chen, L.; Zhang, L.; Liao, Y.; Aizpurua, J.; Luo, Y.E.; et al. Chemical Mapping of a Single Molecule by Plasmon-enhanced Raman Scattering. *Nature* **2013**, *498*, 82–86. [CrossRef]
45. Hartman, T.; Wondergem, C.S.; Kumar, N.; van den Berg, A.; Weckhuysen, B.M. Surface- and Tip-enhanced Raman Spectroscopy in Catalysis. *J. Phys. Chem. Lett.* **2016**, *7*, 1570–1584. [CrossRef]
46. Sambur, J.B.; Chen, P. Approaches to Single-nanoparticle Catalysis. *Annu. Rev. Phys. Chem.* **2014**, *65*, 395–422. [CrossRef]
47. Yu, Y.; Wijesekara, K.D.; Xi, X.; Willets, K.A. Quantifying Wavelength-dependent Plasmonic Hot Carrier Energy Distributions at Metal/Semiconductor Interfaces. *ACS Nano* **2019**, *13*, 3629–3637. [CrossRef]
48. De Cremer, G.; Sels, B.F.; De Vos, D.E.; Hofkens, J.; Roeffaers, M.B.J. Fluorescence Micro (Spectro) Scopy as a Tool to Study Catalytic Materials in Action. *Chem. Soc. Rev.* **2010**, *39*, 4703–4717. [CrossRef]
49. Chen, P.; Zhou, X.; Andoy, N.M.; Han, K.S.; Choudhary, E.; Zou, N.; Chen, G.; Shen, H. Spatiotemporal Catalytic Dynamics Within Single Nanocatalysts Revealed by Single-molecule Microscopy. *Chem. Soc. Rev.* **2014**, *43*, 1107–1117. [CrossRef]
50. Schlücker, S. Surface-enhanced Raman Spectroscopy: Concepts and Chemical Applications. *Angew. Chem. Int. Ed.* **2014**, *53*, 4756–4795. [CrossRef]
51. Cui, L.; Ren, X.; Yang, X.; Wang, P.; Qu, Y.; Liang, W.; Sun, M. Plasmon-driven Catalysis in Aqueous Solutions Probed by SERS Spectroscopy. *J. Raman Spectrosc.* **2016**, *47*, 877–883. [CrossRef]
52. Chen, T.; Dong, B.; Chen, K.; Zhao, F.; Cheng, X.; Ma, C.; Lee, S.; Zhang, P.; Kang, S.H.; Ha, J.W.; et al. Optical Super-resolution Imaging of Surface Reactions. *Chem. Rev.* **2017**, *117*, 7510–7537. [CrossRef] [PubMed]
53. Wilson, A.J.; Devasia, D.; Jain, P.K. Nanoscale Optical Imaging in Chemistry. *Chem. Soc. Rev.* **2020**, *49*, 6087–6112. [CrossRef] [PubMed]
54. Bohren, C.F.; Huffman, D.R. *Absorption and Scattering of Light by Small Particles*; John Wiley & Sons: New York, NY, USA, 2008.
55. Johnson, P.B.; Christy, R.W. Optical Constants of the Noble Metals. *Phys. Rev. B* **1972**, *6*, 4370. [CrossRef]

56. Yang, H.U.; D'Archangel, J.; Sundheimer, M.L.; Tucker, E.; Boreman, G.D.; Raschke, M.B. Optical Dielectric Function of Silver. *Phys. Rev. B* **2015**, *91*, 235137. [CrossRef]
57. Mie, G. Beiträge zur Optik trüber Medien, speziell kolloidaler Metallösungen. *Ann. Der Phys.* **1908**, *330*, 377–445. [CrossRef]
58. Sousa-Castillo, A.; Comesaña Hermo, M.; Rodriguez-Gonzalez, B.; Pérez-Lorenzo, M.; Wang, Z.; Kong, X.T.; Govorov, A.O.; Correa-Duarte, M.A. Boosting Hot Electron-driven Photocatalysis Through Anisotropic Plasmonic Nanoparticles with Hot Spots in Au–TiO2 Nanoarchitectures. *J. Phys. Chem. C* **2016**, *120*, 11690–11699. [CrossRef]
59. Atta, S.; Pennington, A.M.; Celik, F.E.; Fabris, L. TiO2 on Gold Nanostars Enhances Photocatalytic Water Reduction in the Near-infrared Regime. *Chem* **2018**, *4*, 2140–2153. [CrossRef]
60. Yuan, L.; Lou, M.; Clark, B.D.; Lou, M.; Zhou, L.; Tian, S.; Jacobson, C.R.; Nordlander, P.; Halas, N.J. Morphology-Dependent Reactivity of a Plasmonic Photocatalyst. *ACS Nano* **2020**, *14*, 12054–12063. [CrossRef]
61. Wertz, E.; Isaacoff, B.P.; Flynn, J.D.; Biteen, J.S. Single-molecule Super-resolution Microscopy Reveals How Light Couples to a Plasmonic Nanoantenna on the Nanometer Scale. *Nano Lett.* **2015**, *15*, 2662–2670. [CrossRef]
62. Wertz, E.A.; Isaacoff, B.P.; Biteen, J.S. Wavelength-dependent Super-resolution Images of Dye Molecules Coupled to Plasmonic Nanotriangles. *ACS Photonics* **2016**, *3*, 1733–1740. [CrossRef]
63. Langille, M.R.; Personick, M.L.; Zhang, J.; Mirkin, C.A. Defining Rules for the Shape Evolution of Gold Nanoparticles. *J. Am. Chem. Soc.* **2012**, *134*, 14542–14554. [CrossRef] [PubMed]
64. Lohse, S.E.; Burrows, N.D.; Scarabelli, L.; Liz-Marzán, L.M.; Murphy, C.J. Anisotropic Noble Metal Nanocrystal Growth: The Role of Halides. *Chem. Mater.* **2014**, *26*, 34–43. [CrossRef]
65. Bernardi, M.; Mustafa, J.; Neaton, J.B.; Louie, S.G. Theory and Computation of Hot Carriers Generated by Surface Plasmon Polaritons in Noble Metals. *Nat. Commun.* **2015**, *6*, 1–9. [CrossRef]
66. Brown, A.M.; Sundararaman, R.; Narang, P.; Goddard, W.A., III; Atwater, H.A. Nonradiative Plasmon Decay and Hot Carrier Dynamics: Effects of Phonons, Surfaces, and Geometry. *ACS Nano* **2016**, *10*, 957–966. [CrossRef]
67. Khurgin, J.B. Hot Carriers Generated by Plasmons: Where Are They Generated and Where Do They Go from There? *Faraday Discuss.* **2019**, *214*, 35–58. [CrossRef]
68. Baffou, G.; Quidant, R. Thermo-plasmonics: Using Metallic Nanostructures as Nano-sources of Heat. *Laser Photonics Rev.* **2013**, *7*, 171–187. [CrossRef]
69. Christopher, P.; Xin, H.; Marimuthu, A.; Linic, S. Singular Characteristics and Unique Chemical Bond Activation Mechanisms of Photocatalytic Reactions on Plasmonic Nanostructures. *Nat. Mater.* **2012**, *11*, 1044–1050. [CrossRef]
70. Li, K.; Hogan, N.J.; Kale, M.J.; Halas, N.J.; Nordlander, P.; Christopher, P. Balancing Near-field Enhancement, Absorption, and Scattering for Effective Antenna–reactor Plasmonic Photocatalysis. *Nano Lett.* **2017**, *17*, 3710–3717. [CrossRef]
71. Seemala, B.; Therrien, A.J.; Lou, M.; Li, K.; Finzel, J.P.; Qi, J.; Nordlander, P.; Christopher, P. Plasmon-mediated Catalytic O2 Dissociation on Ag Nanostructures: Hot Electrons or Near Fields? *ACS Energy Lett.* **2019**, *4*, 1803–1809. [CrossRef]
72. Zhou, L.; Martirez, J.M.P.; Finzel, J.; Zhang, C.; Swearer, D.F.; Tian, S.; Robatjazi, H.; Lou, M.; Dong, L.; Henderson, L.; et al. Light-driven Methane Dry Reforming with Single Atomic Site Antenna-reactor Plasmonic Photocatalysts. *Nat. Energy* **2020**, *5*, 61–70. [CrossRef]
73. Ferry, V.E.; Munday, J.N.; Atwater, H.A. Design Considerations for Plasmonic Photovoltaics. *Adv. Mater.* **2010**, *22*, 4794–4808. [CrossRef] [PubMed]
74. Wang, Z.; Dong, Z.; Gu, Y.; Chang, Y.H.; Zhang, L.; Li, L.J.; Zhao, W.; Eda, G.; Zhang, W.; Grinblat, G.; et al. Giant Photoluminescence Enhancement in Tungsten-diselenide–gold Plasmonic Hybrid Structures. *Nat. Commun.* **2016**, *7*, 1–8. [CrossRef] [PubMed]
75. Kneipp, K.; Moskovits, M.; Kneipp, H. *Surface-Enhanced Raman Scattering: Physics and Applications*; Springer Science & Business Media, Berlin, Germany, 2006; Volume 103.
76. Zrimsek, A.B.; Chiang, N.; Mattei, M.; Zaleski, S.; McAnally, M.O.; Chapman, C.T.; Henry, A.I.; Schatz, G.C.; Van Duyne, R.P. Single-molecule Chemistry with Surface-and Tip-enhanced Raman Spectroscopy. *Chem. Rev.* **2017**, *117*, 7583–7613. [CrossRef] [PubMed]

77. Ding, S.Y.; You, E.M.; Tian, Z.Q.; Moskovits, M. Electromagnetic Theories of Surface-enhanced Raman Spectroscopy. *Chem. Soc. Rev.* **2017**, *46*, 4042–4076. [CrossRef] [PubMed]
78. Pilot, R.; Signorini, R.; Durante, C.; Orian, L.; Bhamidipati, M.; Fabris, L. A Review on Surface-enhanced Raman Scattering. *Biosensors* **2019**, *9*, 57. [CrossRef] [PubMed]
79. Le Ru, E.C.; Blackie, E.; Meyer, M.; Etchegoin, P.G. Surface Enhanced Raman Scattering Enhancement Factors: A Comprehensive Study. *J. Phys. Chem. C* **2007**, *111*, 13794–13803. [CrossRef]
80. Kang, L.; Xu, P.; Zhang, B.; Tsai, H.; Han, X.; Wang, H.L. Laser Wavelength-and Power-dependent Plasmon-driven Chemical Reactions Monitored Using Single Particle Surface Enhanced Raman Spectroscopy. *Chem. Commun.* **2013**, *49*, 3389–3391. [CrossRef]
81. Sharma, B.; Frontiera, R.R.; Henry, A.I.; Ringe, E.; Van Duyne, R.P. SERS: Materials, applications, and the future. *Mater. Today* **2012**, *15*, 16–25. [CrossRef]
82. Zhang, H.; Duan, S.; Radjenovic, P.M.; Tian, Z.Q.; Li, J.F. Core–Shell Nanostructure-Enhanced Raman Spectroscopy for Surface Catalysis. *Acc. Chem. Res.* **2020**, *53*, 729–739. [CrossRef]
83. Cardinal, M.F.; Vander Ende, E.; Hackler, R.A.; McAnally, M.O.; Stair, P.C.; Schatz, G.C.; Van Duyne, R.P. Expanding applications of SERS through versatile nanomaterials engineering. *Chem. Soc. Rev.* **2017**, *46*, 3886–3903. [CrossRef] [PubMed]
84. Choi, H.K.; Park, W.H.; Park, C.G.; Shin, H.H.; Lee, K.S.; Kim, Z.H. Metal-catalyzed Chemical Reaction of Single Molecules Directly Probed by Vibrational Spectroscopy. *J. Am. Chem. Soc.* **2016**, *138*, 4673–4684. [CrossRef]
85. de Nijs, B.; Benz, F.; Barrow, S.J.; Sigle, D.O.; Chikkaraddy, R.; Palma, A.; Carnegie, C.; Kamp, M.; Sundararaman, R.; Narang, P.; et al. Plasmonic Tunnel Junctions for Single-molecule Redox Chemistry. *Nat. Commun.* **2017**, *8*, 1–8. [CrossRef] [PubMed]
86. Benz, F.; Schmidt, M.K.; Dreismann, A.; Chikkaraddy, R.; Zhang, Y.; Demetriadou, A.; Carnegie, C.; Ohadi, H.; De Nijs, B.; Esteban, R.; et al. Single-molecule Optomechanics in "Picocavities". *Science* **2016**, *354*, 726–729. [CrossRef] [PubMed]
87. Le Ru, E.C.; Etchegoin, P.G. Single-molecule Surface-enhanced Raman Spectroscopy. *Annu. Rev. Phys. Chem.* **2012**, *63*, 65–87. [CrossRef] [PubMed]
88. Choi, H.K.; Lee, K.S.; Shin, H.H.; Koo, J.J.; Yeon, G.J.; Kim, Z.H. Single-Molecule Surface-Enhanced Raman Scattering as a Probe of Single-Molecule Surface Reactions: Promises and Current Challenges. *Acc. Chem. Res.* **2019**, *52*, 3008–3017. [CrossRef] [PubMed]
89. Stranahan, S.M.; Willets, K.A. Super-resolution Optical Imaging of Single-molecule SERS Hot Spots. *Nano Lett.* **2010**, *10*, 3777–3784. [CrossRef] [PubMed]
90. Chernyshova, I.V.; Somasundaran, P.; Ponnurangam, S. On the Origin of the Elusive First Intermediate of CO_2 Electroreduction. *Proc. Natl. Acad. Sci. USA* **2018**, *115*, E9261–E9270. [CrossRef]
91. Lee, H.K.; Lee, Y.H.; Morabito, J.V.; Liu, Y.; Koh, C.S.L.; Phang, I.Y.; Pedireddy, S.; Han, X.; Chou, L.Y.; Tsung, C.K.; et al. Driving CO_2 to a Quasi-Condensed Phase at the Interface between a Nanoparticle Surface and a Metal–Organic Framework at 1 bar and 298 K. *J. Am. Chem. Soc.* **2017**, *139*, 11513–11518. [CrossRef]
92. Kim, H.; Kosuda, K.M.; Van Duyne, R.P.; Stair, P.C. Resonance Raman and Surface-and Tip-enhanced Raman Spectroscopy Methods to Study Solid Catalysts and Heterogeneous Catalytic Reactions. *Chem. Soc. Rev.* **2010**, *39*, 4820–4844. [CrossRef]
93. Kumari, G.; Zhang, X.; Devasia, D.; Heo, J.; Jain, P.K. Watching Visible Light-driven CO_2 Reduction on a Plasmonic Nanoparticle Catalyst. *ACS Nano* **2018**, *12*, 8330–8340. [CrossRef] [PubMed]
94. Fredriksson, H.; Alaverdyan, Y.; Dmitriev, A.; Langhammer, C.; Sutherland, D.S.; Zäch, M.; Kasemo, B. Hole–mask Colloidal Lithography. *Adv. Mater.* **2007**, *19*, 4297–4302. [CrossRef]
95. Olson, J.; Dominguez-Medina, S.; Hoggard, A.; Wang, L.Y.; Chang, W.S.; Link, S. Optical Characterization of Single Plasmonic Nanoparticles. *Chem. Soc. Rev.* **2015**, *44*, 40–57. [CrossRef] [PubMed]
96. Zhang, Z.; Deckert-Gaudig, T.; Deckert, V. Label-free Monitoring of Plasmonic Catalysis on the Nanoscale. *Analyst* **2015**, *140*, 4325–4335. [CrossRef]
97. Herves, P.; Pérez-Lorenzo, M.; Liz-Marzán, L.M.; Dzubiella, J.; Lu, Y.; Ballauff, M. Catalysis by metallic nanoparticles in aqueous solution: Model reactions. *Chem. Soc. Rev.* **2012**, *41*, 5577–5587. [CrossRef]
98. Kang, L.; Han, X.; Chu, J.; Xiong, J.; He, X.; Wang, H.L.; Xu, P. In Situ Surface-Enhanced Raman Spectroscopy Study of Plasmon-Driven Catalytic Reactions of 4-Nitrothiophenol under a Controlled Atmosphere. *ChemCatChem* **2015**, *7*, 1004–1010. [CrossRef]

99. Keller, E.L.; Frontiera, R.R. Ultrafast Nanoscale Raman Thermometry Proves Heating Is Not a Primary Mechanism for Plasmon-driven Photocatalysis. *ACS Nano* **2018**, *12*, 5848–5855. [CrossRef]
100. Wu, Y.; Yang, M.; Ueltschi, T.W.; Mosquera, M.A.; Chen, Z.; Schatz, G.C.; Van Duyne, R.P. SERS study of the mechanism of plasmon-driven hot electron transfer between gold nanoparticles and PCBM. *J. Phys. Chem. C* **2019**, *123*, 29908–29915. [CrossRef]
101. Pozzi, E.A.; Zrimsek, A.B.; Lethiec, C.M.; Schatz, G.C.; Hersam, M.C.; Van Duyne, R.P. Evaluating Single-molecule Stokes and Anti-stokes SERS for Nanoscale Thermometry. *J. Phys. Chem. C* **2015**, *119*, 21116–21124. [CrossRef]
102. Airy, G.B. On the Diffraction of an Object-glass with Circular Aperture. *Trans. Camb. Philos. Soc.* **1835**, *5*, 283.
103. Thompson, R.E.; Larson, D.R.; Webb, W.W. Precise Nanometer Localization Analysis for Individual Fluorescent Probes. *Biophys. J.* **2002**, *82*, 2775–2783. [CrossRef]
104. Mortensen, K.I.; Churchman, L.S.; Spudich, J.A.; Flyvbjerg, H. Optimized Localization Analysis for Single-molecule Tracking and Super-resolution Microscopy. *Nat. Methods* **2010**, *7*, 377–381. [CrossRef] [PubMed]
105. Betzig, E.; Patterson, G.H.; Sougrat, R.; Lindwasser, O.W.; Olenych, S.; Bonifacino, J.S.; Davidson, M.W.; Lippincott-Schwartz, J.; Hess, H.F. Imaging Intracellular Fluorescent Proteins at Nanometer Resolution. *Science* **2006**, *313*, 1642–1645. [CrossRef] [PubMed]
106. Hess, S.T.; Girirajan, T.P.K.; Mason, M.D. Ultra-high Resolution Imaging by Fluorescence Photoactivation Localization Microscopy. *Biophys. J.* **2006**, *91*, 4258–4272. [CrossRef]
107. Rust, M.J.; Bates, M.; Zhuang, X. Sub-diffraction-limit Imaging by Stochastic Optical Reconstruction Microscopy (STORM). *Nat. Methods* **2006**, *3*, 793–796. [CrossRef]
108. Johlin, E.; Solari, J.; Mann, S.A.; Wang, J.; Shimizu, T.S.; Garnett, E.C. Super-resolution Imaging of Light–matter Interactions near Single Semiconductor Nanowires. *Nat. Commun.* **2016**, *7*, 1–6. [CrossRef]
109. Mack, D.L.; Cortés, E.; Giannini, V.; Török, P.; Roschuk, T.; Maier, S.A. Decoupling Absorption and Emission Processes in Super-resolution Localization of Emitters in a Plasmonic Hotspot. *Nat. Commun.* **2017**, *8*, 1–10. [CrossRef]
110. Hamans, R.F.; Parente, M.; Castellanos, G.W.; Ramezani, M.; Gómez Rivas, J.; Baldi, A. Super-resolution Mapping of Enhanced Emission by Collective Plasmonic Resonances. *ACS Nano* **2019**, *13*, 4514–4521. [CrossRef]
111. Ristanović, Z.; Kubarev, A.V.; Hofkens, J.; Roeffaers, M.B.; Weckhuysen, B.M. Single Molecule Nanospectroscopy Visualizes Proton-transfer Processes Within a Zeolite Crystal. *J. Am. Chem. Soc.* **2016**, *138*, 13586–13596. [CrossRef]
112. Roeffaers, M.B.J.; Sels, B.F.; Uji-i, H.; De Schryver, F.C.; Jacobs, P.A.; De Vos, D.E.; Hofkens, J. Spatially Resolved Observation of Crystal-face-dependent Catalysis by Single Turnover Counting. *Nature* **2006**, *439*, 572–575. [CrossRef]
113. Martin-Fernandez, M.L.; Tynan, C.J.; Webb, S.E.D. A 'Pocket Guide' to Total Internal Reflection Fluorescence. *J. Microsc.* **2013**, *252*, 16–22. [CrossRef] [PubMed]
114. Ha, J.W.; Ruberu, T.P.A.; Han, R.; Dong, B.; Vela, J.; Fang, N. Super-resolution Mapping of Photogenerated Electron and Hole Separation in Single Metal–semiconductor Nanocatalysts. *J. Am. Chem. Soc.* **2014**, *136*, 1398–1408. [CrossRef] [PubMed]
115. Zhou, X.; Xu, W.; Liu, G.; Panda, D.; Chen, P. Size-dependent Catalytic Activity and Dynamics of Gold Nanoparticles at the Single-molecule Level. *J. Am. Chem. Soc.* **2010**, *132*, 138–146. [CrossRef] [PubMed]
116. Gulati, A.; Liao, H.; Hafner, J.H. Monitoring Gold Nanorod Synthesis by Localized Surface Plasmon Resonance. *J. Phys. Chem. B* **2006**, *110*, 22323–22327. [CrossRef]
117. Zou, N.; Zhou, X.; Chen, G.; Andoy, N.M.; Jung, W.; Liu, G.; Chen, P. Cooperative Communication within and between Single Nanocatalysts. *Nat. Chem.* **2018**, *10*, 607–614. [CrossRef]
118. Zou, N.; Chen, G.; Mao, X.; Shen, H.; Choudhary, E.; Zhou, X.; Chen, P. Imaging Catalytic Hotspots on Single Plasmonic Nanostructures via Correlated Super-resolution and Electron Microscopy. *ACS Nano* **2018**, *12*, 5570–5579. [CrossRef]
119. Li, W.; Miao, J.; Peng, T.; Lv, H.; Wang, J.G.; Li, K.; Zhu, Y.; Li, D. Single-molecular Catalysis Identifying Activation Energy of the Intermediate Product and Rate-limiting Step in Plasmonic Photocatalysis. *Nano Lett.* **2020**, *20*, 2507–2513. [CrossRef]

120. Chen, T.; Zhang, Y.; Xu, W. Single-molecule Nanocatalysis Reveals Catalytic Activation Energy of Single Nanocatalysts. *J. Am. Chem. Soc.* **2016**, *138*, 12414–12421. [CrossRef]
121. Liu, X.; Chen, T.; Xu, W. Revealing the Thermodynamics of Individual Catalytic Steps Based on Temperature-dependent Single-particle Nanocatalysis. *Phys. Chem. Chem. Phys.* **2019**, *21*, 21806–21813. [CrossRef]
122. Mao, X.; Liu, C.; Hesari, M.; Zou, N.; Chen, P. Super-resolution Imaging of Non-fluorescent Reactions via Competition. *Nat. Chem.* **2019**, *11*, 687–694. [CrossRef]
123. Chen, G.; Zou, N.; Chen, B.; Sambur, J.B.; Choudhary, E.; Chen, P. Bimetallic Effect of Single Nanocatalysts Visualized by Super-resolution Catalysis Imaging. *ACS Cent. Sci.* **2017**, *3*, 1189–1197. [CrossRef] [PubMed]
124. Ebrahimi, S.; Akhlaghi, Y.; Kompany-Zareh, M.; Rinnan, Å. Nucleic Acid Based Fluorescent Nanothermometers. *ACS Nano* **2014**, *8*, 10372–10382. [CrossRef] [PubMed]
125. Carattino, A.; Caldarola, M.; Orrit, M. Gold Nanoparticles as Absolute Nanothermometers. *Nano Lett.* **2018**, *18*, 874–880. [CrossRef] [PubMed]
126. Kamarudheen, R.; Kumari, G.; Baldi, A. Plasmon-driven Synthesis of Individual Metal@Semiconductor Core@Shell Nanoparticles. *Nat. Commun.* **2020**, *11*, 1–10. [CrossRef] [PubMed]
127. Baffou, G. *Thermoplasmonics: Heating Metal Nanoparticles Using Light*; Cambridge University Press: Cambridge, UK, 2017.
128. Barella, M.; Violi, I.L.; Gargiulo, J.; Martinez, L.P.; Goschin, F.; Guglielmotti, V.; Pallarola, D.; Schlücker, S.; Pilo-Pais, M.; Acuna, G.P.; et al. In Situ Photothermal Response of Single Gold Nanoparticles through Hyperspectral Imaging Anti-Stokes Thermometry. *ACS Nano* **2020**. [CrossRef] [PubMed]
129. Baffou, G.; Bordacchini, I.; Baldi, A.; Quidant, R. Simple Experimental Procedures to Distinguish Photothermal from Hot-carrier Processes in Plasmonics. *Light. Sci. Appl.* **2020**, *9*, 1–16. [CrossRef] [PubMed]
130. Félidj, N.; Aubard, J.; Lévi, G.; Krenn, J.R.; Salerno, M.; Schider, G.; Lamprecht, B.; Leitner, A.; Aussenegg, F. Controlling the Optical Response of Regular Arrays of Gold Particles for Surface-enhanced Raman Scattering. *Phys. Rev. B* **2002**, *65*, 075419. [CrossRef]
131. Stewart, M.E.; Anderton, C.R.; Thompson, L.B.; Maria, J.; Gray, S.K.; Rogers, J.A.; Nuzzo, R.G. Nanostructured Plasmonic Sensors. *Chem. Rev.* **2008**, *108*, 494–521. [CrossRef]
132. Fan, M.; Andrade, G.F.; Brolo, A.G. A Review on the Fabrication of Substrates for Surface Enhanced Raman Spectroscopy and Their Applications in Analytical Chemistry. *Anal. Chim. Acta* **2011**, *693*, 7–25. [CrossRef]
133. Cortés, E.; Xie, W.; Cambiasso, J.; Jermyn, A.S.; Sundararaman, R.; Narang, P.; Schlücker, S.; Maier, S.A. Plasmonic Hot Electron Transport Drives Nano-localized Chemistry. *Nat. Commun.* **2017**, *8*, 1–10. [CrossRef]
134. Jermyn, A.S.; Tagliabue, G.; Atwater, H.A.; Goddard, W.A., III; Narang, P.; Sundararaman, R. Transport of Hot Carriers in Plasmonic Nanostructures. *Phys. Rev. Mater.* **2019**, *3*, 075201. [CrossRef]
135. Baffou, G.; Quidant, R.; García de Abajo, F.J. Nanoscale Control of Optical Heating in Complex Plasmonic Systems. *ACS Nano* **2010**, *4*, 709–716. [CrossRef] [PubMed]

Publisher's Note: MDPI stays neutral with regard to jurisdictional claims in published maps and institutional affiliations.

 © 2020 by the authors. Licensee MDPI, Basel, Switzerland. This article is an open access article distributed under the terms and conditions of the Creative Commons Attribution (CC BY) license (http://creativecommons.org/licenses/by/4.0/).

MDPI
St. Alban-Anlage 66
4052 Basel
Switzerland
Tel. +41 61 683 77 34
Fax +41 61 302 89 18
www.mdpi.com

Nanomaterials Editorial Office
E-mail: nanomaterials@mdpi.com
www.mdpi.com/journal/nanomaterials

www.ingramcontent.com/pod-product-compliance
Lightning Source LLC
LaVergne TN
LVHW070626100526
838202LV00012B/740